Brain Culture

Brain Culture

Neuroscience and Popular Media

DAVI JOHNSON THORNTON

RUTGERS UNIVERSITY PRESS

NEW BRUNSWICK, NEW JERSEY, AND LONDON

LIBRARY OF CONGRESS CATALOGING-IN-PUBLICATION DATA

Thornton, Davi Johnson, 1978–
    Brain culture : neuroscience and popular media / Davi Johnson Thornton.
        p. ; cm.
    Includes bibliographical references and index.
    ISBN 978–0–8135–5012–1 (hardcover : alk. paper) — ISBN 978–0–8135–5013–8
(pbk. : alk. paper)
    1. Neurosciences.   2. Science in popular culture.   3. Science in mass media.
I. Title.
    RC343.T563 2011
    616.8—dc22   2010035284

A British Cataloging-in-Publication record for this book is available from the
British Library.

Visit our Web site: http://rutgerspress.rutgers.edu

Manufactured in the United States of America

CONTENTS

ACKNOWLEDGMENTS

Since I first began researching popular representations of brain imaging, I have been supported and encouraged by many friends and colleagues who have generously engaged in conversation, commented on early drafts and presentations, and generally contributed to my thinking and writing in a number of different ways. I would especially like to thank Kevin DeLuca for his friendship, guidance, and encouragement. I would also like to thank Celeste Condit, Tom Lessl, Beth Preston, and Christine Harold for their patience and generosity. I would also like to thank my reading group colleagues, Nate Stormer, Majia Nadesan, and Megan Foley, whose conversations have pushed my own thinking forward on a number of fronts. I am especially grateful to Michelle Gibbons for reading sections of this manuscript and providing helpful feedback. There are a number of other colleagues who have influenced and inspired me throughout my academic career, and in particular I thank John Sloop, Mike Janas, and Christina Morus. In addition, I owe much to my colleagues at Southwestern for all the different ways they have encouraged and supported me throughout this project. I would especially like to thank Bob Bednar, Julia Johnson, David Olson, Elizabeth Green Musselman, Eric Selbin, Mary Grace Neville, Erika Berroth, Eileen Russell, and Katy Ross. Finally, I would like to thank my family for their support—my parents, David and Gayle Johnson; my brothers, Taylor, Graham, Austin, and Brett; my favorite sister-in-law, Mindy; and especially my husband, Paul Thornton.

In addition to my friends and colleagues, I thank Leslie Mitchner and the staff at Rutgers for their patience and commitment to this book. I am also grateful to Javier Mariscal and Estudio Mariscal for allowing me to use "New Worlds" as cover art. I am also grateful for Southwestern University's faculty development programs, which have aided me in completing this volume.

A different version of chapter 3 was previously published by Springer Science + Business Media as "'How Do You Know Unless You Look?': Brain Imaging, Biopower, and Practical Neuroscience," *Journal of Medical Humanities* 29, no. 3 (2008): 147–161. A previous version of chapter 5 was published by Routledge as "Psychiatric Power: The Post-Museum as a Site of Rhetorical Alignment," *Communication and Critical/Cultural Studies* 5, no. 4 (2008): 344–362.

Brain Culture

1

The Rhetorical Brain

In 2008, the neuroscientist Ryuta Kawashima's brain training manual *Train Your Brain More: Better Brainpower, Better Memory, Better Creativity* hit the book stands, joining an already impressive array of books promising readers that by working on their brains, they can achieve their greatest desire, whether that be enhanced memory, weight loss, a better job, or simply a more peaceful life. *Train Your Brain More* is just one of dozens of books published in the past decade that provide readers with concrete, practical instructions for "rewiring" and "training" their brains in order to transform their lives. These books, with titles like *Rewire Your Brain: Think Your Way to a Better Life*; *The Brain Advantage: Become a More Effective Business Leader Using the Latest Brain Research*; and *Retrain Your Brain, Reshape Your Body: The Breakthrough Brain-Changing Weight-Loss Program*, use brain imaging research to visualize the brain as an array of dynamic processes that can be willfully changed, rather than an inert organ with permanently fixed structures and functions.[1] In *Train Your Brain More*, Kawashima illustrates the plasticity, or malleability, of the brain with brightly colored brain images that graphically depict the positive effects of his brain training regimen. Laid out in workbook fashion, Kawashima's program takes readers through a series of brain "exercises" designed to improve mental function and overall performance in life. The book includes a detailed chart that instructs readers to carefully track their exercise routines on a day-to-day basis, documenting improvements in brain health over the course of the program. According to Kawashima, if practiced consistently, these training exercises will lower one's "brain age" by increasing both the strength and number of connections among neurons in the brain.

What is fascinating about Kawashima's book and others like it is that they frame brain imaging and neuroscience as an accessible body of knowledge that has direct, concrete applications for almost every area of daily life.

According to this popular neuroscience, because the brain is the source of literally all human thought, emotion, and behavior, willful efforts to improve the brain will naturally lead to superior intelligence, greater emotional stability, and improved performance in the home, at the gym, and in the workplace. In slightly different terms, if one wants to become better in some aspect of life—smarter, fitter, more sociable, or more competitive—the key to improvement is always the brain, regardless of the specific type of improvement desired. Want a better body? Work on your brain. Want better children? Work on your brain (and theirs). Want a better job? Work on your brain. In this context, having a healthy brain is not simply a matter of avoiding injuries and illnesses, but rather is tied to endless projects of self-optimization in which individuals are responsible for continuously working on their own brains to produce themselves as better parents, workers, and citizens.

Brain Culture is a study of brain science at the level of culture, or in terms of its impact on the practices of everyday life. This book reveals the means by which Americans have come to define themselves and their societies through this brain-based healthism. What are the rhetorical or discursive processes through which individuals come to see themselves as "neurochemical selves," managing their lives by working and acting for the health of the brain?[2] How have we come to understand the brain as something that we must attend to and care for through scientifically validated personal regimens and elaborate social policies? What rhetorical processes are at work when public policy discourse is saturated with "brain talk"? These are the questions that propel this inquiry into the rhetorical brain. Through a series of interconnected case studies, I trace the circulation of the rhetorical brain through diverse social contexts, attending to ways in which public discourse about the brain and contemporary technologies that visualize the brain mutually influence and condition one another.

The brain training phenomenon is just one small part of our fascination with all things brain, a fixation that extends far beyond the spheres of science and medicine and infiltrates virtually every corner of daily existence. Neuroscience and brain imaging are being used to define a number of issues at all levels of life—from personal improvement to public policy. The brain has become, in Kawashima's words, not merely an individual obsession but also "an important social priority."[3] Kawashima's own research is a case in point. In 2001, several years before he delved into the brain training market, Kawashima pioneered a highly publicized study that used brain images to show that when children play video games, the frontal lobes of their brains are rendered less active, making it difficult for them to control their antisocial and aggressive impulses. Kawashima's warning was stark. At a learning conference in Britain, he announced, "The implications are very serious for an increasingly violent society and these students will be doing more and more bad things if they are playing video games and not doing other things like reading aloud or learning arithmetic."[4]

By itself, Kawashima's link between video games and violence was nothing new—for decades, critics have accused video games of breeding antisocial behaviors and attitudes. What distinguished Kawashima's claims and contributed to escalating public panic over video games was his focus on the brain. Kawashima articulated the link between video games and violence in terms of biological processes in the brain and provided compelling visual evidence in the form of dramatic brain images vividly depicting frontal lobe deficiencies in children. Other rationales commonly used to link video games to violence, such as the much-contested argument that children mimic the violence they witness on-screen, proved much less convincing than the authoritative neuroscientific claims backed by visual data. In 2005, brain imaging research documenting a link between video games and frontal lobe deficiencies was so persuasive that it was used by public officials in Illinois to fight for passage of the Safe Games Illinois Act, legislation requiring extensive warning labels on certain games and severe penalties for the sale of selected games to minors.[5] Although the act was later struck down, the extent to which the video game controversy of the early and mid-2000s focused attention on brain images and the role of the frontal lobes in human behavior exemplifies the ways that enduring social concerns, including child rearing, commerce, social stability, crime, and the prevention of violence, are increasingly defined and debated with reference to the brain.

Visualizing the Rhetorical Brain

As the examples of the video game controversy and the brain training phenomenon show, the brain's centrality in cultural life is largely attributed to persuasive brain images that circulate widely, from the pages of self-improvement manuals to courtroom testimony. A key feature of the rhetorical brain (a term I use to describe the unique ways we conceptualize the brain in contemporary culture) is that we can see it—and its visual characteristics are crucial to the way that we understand it. The brightly colored scans of brain function produced by digital imaging technologies, including PET scans (positron-emission tomography), fMRI (functional magnetic resonance imaging), EEG (electroencephalography), and MEG (magnetoencephalography), are ubiquitous features of the contemporary media landscape. These widely disseminated brain images are used to support a host of claims about the role of the brain as both the source and outcome, or cause and effect, of all human activity. Brain training manuals, for example, use brain images to support claims that willful efforts can have an immediate, visible effect on brain function. At the same time, however, these books often describe the brain as the cause of all of our successes and failures. By training the brain, success—whether it is defined as a better job, superior intelligence, or a fit body—is guaranteed because the brain is precisely what is responsible for producing good performance in all these areas. In these contexts, brain images are put to dual uses—they are often used to substantiate biological determinism ("your brain is responsible for everything you do"), but they are also deployed to support claims that emphasize individual agency and responsibility ("you must take specific actions to guarantee the health of your brain").

There is, then, something of a paradox that permeates contemporary brain culture that is enabled by the flexibility of brain images. Although brain images carry the authoritative weight of science, their meanings are not clear-cut, as they can be used to support competing claims. One task of *Brain Culture* is to map out the implications of this paradox. My objective is not to resolve the tensions between determinism and agency that inhabit popular neuroscience, but instead to trace the diverse appearances of these tensions as they affect individuals and societies. It is in part because of these tensions that the brain obsession has such purchase over our lives. Whether we are trying to account for our behaviors, emotions, and personalities, or seeking an identifiable target for our initiatives to improve ourselves and our societies, popular neuroscience is an accessible language that can accommodate both our desires to assign responsibility ("my brain made me do it") and our desires to confirm the transformative potential of our willful efforts ("I have the power to change my brain").

The ways in which brain images are consumed or interpreted are mediated by the contexts in which they circulate—social, historical, and economic. In and of themselves, digital brain images do not have a fixed or determined meaning. Their rhetorical force arises from the ways they are framed, contextualized, and used in different settings. By tracing the circulation of brain scans and related brain talk, I attend to the interdependent and mutually constitutive nature of texts and contexts, brain scans and sociocultural milieus. When I describe the brain as rhetorical, I mean two things. First, the brain is not simply natural, but also mediated; the ways in which we know, understand, and visualize the brain are historically specific and contingent on available grammars and technologies. Second, the mediated brain exerts social effects, or does things in the world—it has rhetorical force. To say that the brain is mediated is not to say that it is immaterial, or simply a socially constructed fiction. The rhetorical brain circulates throughout society in a number of material forms—images, concepts, colloquialisms, self-help books, museum exhibits—and exerts real-world effects at a number of levels, from individual attitudes and behaviors to public policy. Further, the relationship between the rhetorical brain and broader sociocultural and economic assemblages is not one-way: just as social arrangements are conditioned by the circulation of brain discourses, so conceptualizations of the brain and neuroscience are mediated by the broader contexts in which they function. There is a relationship of mutual conditioning and influence that cannot be reduced to any simple cause-and-effect analysis.

Brain images function as a veridical, or truthful, discourse, invested with the ethos of scientific authority. Yet the brain images are not identical to the "wet brain," the label scientists use to refer to the actual human brain. Brain scan images are digital renderings of graphic data, and the brain they depict looks quite different from an actual wet brain or a representation of one, such as a photograph.[6] Yet it is the digitally rendered brain that has truth-value in our society, not the inert wet brain. It is the digital brain that Steven Johnson describes as "authentic," a "pure vision of the mind's inner life."[7] What factors make brain scan images and neuroscientific terminologies a veridical discourse? To answer this, it is not enough to look at their scientific status; instead, it is necessary to undertake a broader, contextual analysis of the current social arrangements in which these images accrue their force. The question of what theories and images possess epistemic authority in science is not independent from cultural, social, and other extra-scientific factors. Thus the brain is rhetorical not only because it is mediated, but because its very ontology—or what we think that the brain is in essence—is conditioned by the social, political, and economic milieu.

In the context of science, Bruno Latour describes material mediations such as brain scans as "immutable mobiles," visual and verbal inscriptions that circulate as persuasive resources in scientific practice.[8] In the case of brain scans, these immutable mobiles, generated in scientific circles, have persuasive force outside strictly scientific arenas. Brain scan images certainly play a significant role in the generation of scientific data and the formulation of scientific knowledge, but what work do they do in public or popular contexts? In what venues do they circulate, and in what meanings, values, and interpretive frameworks are they embedded? Again, the processes of circulation go both ways: just as the scientific authority of brain scans gives them a particular rhetorical force in popular culture, social and economic contexts shape conceptions of scientific practice and knowledge. A recurring theme is that neither science and "nature" nor society and "nurture" can be taken for granted as clearly demarcated regions of human life.

Calculating the Rhetorical Brain: Assessments and Interventions

If the wide variety of messages associated with the rhetorical brain could be reduced to a single statement, it would be that each individual life, as well as life at the level of the population, is calculable: something that must be carefully assessed and calibrated through technical interventions. Kawashima's concept of brain age is an excellent example of the way life is made calculable through popular discourses of the brain. In the wake of the public clamor over video games spurred by Kawashima's early studies linking games to frontal lobe deficiencies, Nintendo approached Kawashima about creating a video game that would optimize brain function. Kawashima came up with *Brain Age*, an extremely popular video game series first released in 2005 and successfully marketed for widespread use on the Nintendo DS mobile platform. As of 2009, *Brain Age* and *Brain Age 2* had sold more than thirty-one million copies worldwide, and the game is credited with kicking off a global brain fitness craze that includes the development of many similar games and brain training tools.[9] Kawashima's game promises to help consumers keep their brains "young," reducing "brain age" by staving off mental decline that occurs naturally with age, as well as improving brain function for competitive advantage in the workplace ("a video game that helps you outsmart your coworkers").[10] To start *Brain Age*, consumers first undergo a diagnostic test that determines one's brain age as a specific number (many similar brain age tests are now available online from different organizations). From there, individuals have a benchmark against which to measure the effects of their improvement activities. To reduce one's brain age (Nintendo's advertising defines twenty as the ideal brain age), one must consistently play the video game and engage in its specific exercises for brain improvement.

Brain Age shows how popular neuroscience makes life calculable in terms of both assessment and intervention. First, *Brain Age* assigns a concrete number to one's brain health, thereby situating health as something that can be measured and quantified. This number is framed as a more authentic, accurate barometer of life status than other indicators, such as one's natural age. Second, the game promises that brain health can be improved through deliberate, calculated interventions. *Brain Age* exemplifies a theme that runs throughout popular neuroscience: not only is life rendered calculable through the images and discourses of the brain, but individuals have the responsibility for applying these lenses to their own lives. This book looks at the ways that this message is disseminated in popular media. Through various popular media, including self-help books, entertainment television, and advertising, individuals without medical or scientific training are encouraged to take up the languages of neuroscience and use them as accessible vocabularies for articulating their experiences and desires. In these forums, brain science is not only translated for public understanding, but also sold as a key means of self-improvement and social transformation.

The link between popular neuroscience and injunctions for self-improvement is not limited to video games or brain training manuals. For example, in January 2007 *Time* magazine published a special issue on the brain. The issue features an array of articles on the science of the brain, with topics including everything from scientific methods of studying the brain to the effects of stress on brain function to the role of the brain in processing memories. The interesting thing is that scientific facts about the brain are presented as information that makes urgent, practical demands on readers. Readers are instructed that understanding the brain in terms of its scientific function is integral to living a life that maximizes one's true potential. The cover title of the issue, "The Brain: A User's Guide," emphasizes this pragmatic role that brain science plays for each individual. Glossy brain scan images are dispersed throughout the stories, illustrating for readers the centrality of their brains in such everyday activities as focusing at work, relaxing, and sleeping. Scientific terminologies like "frontal, medial, and temporal lobes," "hippocampus," and "cortisol cascade" are littered among colloquialisms such as "stress," "self-esteem," and "mental skills." The scan images "prove" the truthfulness of the scientific terminologies, literally showing readers that what they commonly think of as "self-esteem" or "lack of focus" are in actuality neurological events that can be named with precise, medicoscientific words. By oscillating between colloquial and scientific expression, the latter is suggested as an accessible and highly relevant lexicon to be used for everyday living. One of the articles in the *Time* issue, for instance, uses brain scan evidence to show that happiness and contentment are, in the language of the brain, a pattern of neural activity marked by greater activity in the left prefrontal cortex than the right.[11] This knowledge has practical import because it reveals the importance of *training* the brain through daily practice to create a desirable balance of left and right prefrontal cortex activity.

As "The Brain: A User's Guide" suggests, the dissemination of brain-based vocabularies is a primary means of making everyday life technical, a matter of calculated self-supervision and maintenance. When everyday affairs are translated into the scientific languages of neuroscience, they are endowed with regulatory significance. All those thoughts, emotions, and behaviors that can be connected to the brain are now amenable to rational management. Controlling stress, for example, becomes a matter of maintaining appropriate hippocampus activity, a state that can be achieved by manipulating cortisol levels. The necessary interventions include deep breathing, healthy relationships, proper nutrition and exercise, and adequate sleep. In other words, the means of brain regulation are all in the form of habits of regular living, but they are articulated as technical modes of managing brain biology in the name of self-optimization. This discursive production of neurochemical selves shapes individuals' self-understanding, and it works to persuasively channel individuals' behaviors in ways that have broader cultural, political, and economic significance.

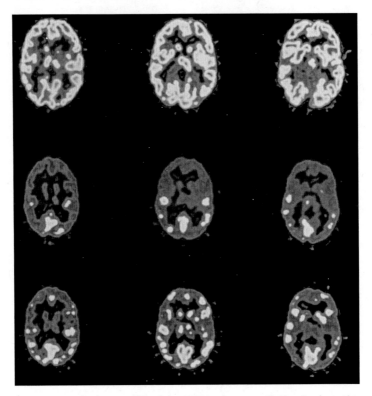

FIGURE 1 Brain images like these PET images correlating brain activity and cocaine use have become familiar features in popular media.

Brookhaven National Laboratory/Getty Images

Even brain training devices directed toward individuals for personal improvement have broader cultural and political significance. For example, skyrocketing sales of brain fitness programs are being fueled not only by individual consumers but also by insurers, employers, and even government agencies, which view these programs as tools to increase worker productivity and reduce health costs.[12] OptumHealth, a group that develops wellness programs for U.S. employers, recently introduced programs to target worker productivity through brain training routines. As Eugene Baker of OptumHealth states, "Improving brain health can result in less presenteeism, the tendency to be at work but be distracted and not able to focus. If you look at disability costs, absenteeism and presenteeism account for most of the medical costs, and that's a good reason for employers to be focused on brain health."[13] In this example, brain health is assigned a measurable, economic value and framed as something that can be improved according to cost-benefit analyses. Also inspired by the idea that the population's brain health can be translated into calculations of economic profitability, insurance companies, including Humana and MetLife, have jumped on the brain fitness bandwagon and initiated programs designed to encourage clients to proactively optimize their brain health.[14] The government, too, is taking an interest in the benefits brain training might provide to the population. For example, the 2005 White House Conference on Aging featured testimony on the importance of brain fitness, highlighting its social and economic benefits in terms of a healthy, productive citizenry.

Brain training, then, is framed as something that is in everyone's interests. Individual consumers purchase *Brain Age* or training manuals because they desire self-improvement; employers invest in brain fitness tools because they ensure worker productivity, lower costs, and greater profits; and government agencies are attracted to programs that promise a happier, healthier citizenry and minimize the financial and regulatory burdens associated with an aging population. Popular neuroscience's ability to facilitate this convergence of interests is a key focus of *Brain Culture*. This field produces a powerful discourse, and it exerts real, sometimes troubling effects in contemporary culture. Yet it is a discourse that rarely works from the top down, such as by imposing restrictive mandates or coercing individuals through forceful dictates. Rather, popular neuroscience is powerful because it is useful and capable of accommodating multiple, even competing interests and desires of individuals, corporations, and public and private authorities. I use the term "governmental" to get at this notion of power that works through coordination and facilitation, in the sense that Foucault describes in his discussions of governmentality.[15] Governmentality refers generally to the ways in which individuals come to regulate themselves in the name of their own good (health, happiness, wealth) in the apparent absence of direct control by either the state or other powerful interests such as corporations. Michel Foucault's most succinct definition of governmentality is "regulation at a distance." The concept is based on the premise that power is not always or even primarily exercised in overt, top-down fashion. For instance, there is not a clearly identifiable central source of power, such as the state, that exercises its rule by commanding or forcing people to behave in certain ways. Instead, power works in more subtle ways. People often come to regulate their own behavior in ways that line up with the interests of other powers (e.g., the state or corporations), even though these authorities appear to have a very "hands-off" approach. In the case of popular neuroscience, for instance, the assessments and interventions disseminated to individuals for their own benefit ("it's in your own best interests to improve your brain health") also govern individuals in the sense that they shape individuals into "good consumers" and "good citizens." For example, when individuals take up biological and psychiatric vocabularies for framing their own lives, these languages often dispose them to actively participate in various initiatives—ranging from the consumption of therapy, including medication, to parenting classes to productive workplace behaviors—all as part of their own pursuit of personal fulfillment and brain health.[16] Thus people are not primarily governed through force or coercion but through rhetorical means, or the dissemination of attractive, useful discourses that channel individuals' attitudes and behaviors in ways that converge with other interests.

Optimization and the Expansion of Health

The dramatic saga of Nintendo DS and Kawashima's *Brain Age* video game illustrates brain culture's intricate confluences of science, entertainment, commerce, and politics. Not surprisingly, Kawashima was a target for criticism when he accepted Nintendo's offer to create a brain-based video game in the wake of public outcry over video games sparked by his own research.[17] Although the story of *Brain Age* has been told as a cautionary tale about corruption, greed, and the corporate seduction of science, I am more interested in what this story tells us about the ways that popular neuroscience is changing our definitions of health and illness, and ultimately how we understand and live our lives. Brain imaging has application far beyond conventional medical uses, and it is used not only to diagnose pathology and identify risks to the healthy brain, but also to determine what technologies (attitudes, behaviors, interventions) can best optimize brain function, improving the abilities of even an apparently healthy brain. In other words, having a normal brain, whatever that might be, is no longer enough: according to Nintendo's webpage for *Brain Age*, deliberate attention to and care for the brain are needed "even if you lead a normal, healthy life."[18] As these advertisements communicate, brain health is not a guaranteed, natural state occasionally damaged by risky exposures or hijacked by pathological intrusions, or even a goal or benchmark against which one can comparatively gauge their own health status; rather, it is a resource that can be acquired without limit. One task of this book is to explore the implications of these changing conceptions that understand health as something to be accumulated rather than as a natural set point or approachable benchmark.

At the same time as the very meanings of what counts as health are changing, health is becoming a dominant discourse for understanding and evaluating multiple aspects of life that extend far beyond medicine's traditional purview. Robert Crawford describes this expansion of health as "healthism," a pervasive health consciousness in which health becomes "a pan-value or standard by which an expanding number of behaviors and social phenomena are judged." Healthism produces "health-governed" identities, individuals who "define their malaise and their goals, along with the implied strategies for alleviation or fulfillment," in the terminologies of health.[19] The language of health is powerful but also ambiguous—what counts as health and what counts as illness are questions as likely to be settled by political and rhetorical means as by clear-cut scientific data, especially in the context of psychiatry. Debates over the meanings of health can have a range of consequences, affecting everything from insurance coverage and social services to individuals' willingness to purchase and consume medications or pursue treatments.

In the early 2000s, the Bush administration engaged with questions of mental health and illness on several different fronts, and its paradoxical dealings with mental health issues illustrate the deeply political nature of questions of health and illness. In October 2003, the President's Council on Bioethics released a report titled *Beyond Therapy: Biotechnology and the Pursuit of Happiness*. In the opening letter, Chairman Leonard Kass pinpoints the focus of the report as the "dual uses" of recent technologies that alter mind and body.[20] The report describes the advance of psychiatric medications, dubbed "mood brighteners" by the council, which are attractive not only to people who are "sick" but also to normal people who simply want to perform better in their daily lives. Kass summarizes the conclusions of the council thus: "We want to be happy—but not because of a drug that gives us happy feelings without the real loves, attachments, and achievements that are essential for true human flourishing."[21] The "medicalization of self-understanding" is bemoaned as a "great advance for biological reductionism against the citadel of mind and soul."[22]

In 2002, the psychiatrist Peter Kramer, author of *Listening to Prozac*, and the bioethicist Carl Elliott testified before the President's Council on Bioethics on a related issue. Kramer and Elliott addressed concerns related to the burgeoning availability of "enhancement technologies," or the use of biotechnologies to make otherwise "healthy" individuals, in Kass's words, "look younger, perform better, feel happier, or become more 'perfect.'"[23] The council heard much testimony on the complexities of the issues involved in the neuroscience revolution, but it tended to take a skeptical view of biotechnologies designed to alter mind and mood. For George W. Bush and the council, recent medical techniques threaten the very authenticity of human existence, tempting society "to settle for a shallow and shrunken imitation."[24]

Despite the deep reservations toward psychiatric enhancement expressed by the council, President Bush avidly supported psychiatric medications in other contexts. In the same year Kramer and Elliott testified before the Council on Bioethics, Bush created the New Freedom Commission on Mental Health, declaring that Americans with mental illness "deserve" both "understanding" and "excellent care."[25] The commission's 2003 report quotes the president saying, "Americans must understand and send this message: mental disability is not a scandal—it is an illness. And like physical illness, it is treatable, especially when the treatment comes early."[26] The report goes on to drive home the message that mental illness is a physical illness, and that the failure to accept this message is to participate in "stigma," a socially and morally unacceptable form of discrimination. The report argues persistently for expansion of mental health care, including making "evidence-based, state-of-the-art medications . . . standard practice."[27]

The tension between these two moral postures toward psychiatric medications—on the one hand, skepticism toward tinkering with the authenticity of human life, and on the other hand, indignation at inadequate access—hinges in part on the question of what counts as health and illness. For persons who are genuinely ill, medications are akin to a right; for persons who are not ill, medications are a dehumanizing corruption of the soul. The difficulty is that the boundaries between health and illness, particularly in the context of mental health and illness, have never been clear-cut. A cursory glance at the history of psychiatry shows that definitions of mental illness are always changing and often contested.[28] What it means to be normal at any given time is heavily dependent on a variety of factors—cultural, economic, political—that are not reducible to evidence-based science.

In our psychiatric culture, this historical fuzziness of the health-versus-illness divide is compounded by the persuasive force of brain imaging technologies that purport to show the biological basis of all aspects of experience. When moods and behaviors can be persuasively linked to brain biology, it is easier to conceptualize them along the lines of pathology. For example, the claim that video games are harmful to brain health is far more credible when brain images visualize the neural correlates of individuals while they are playing video games. Even if fundamental questions about causality are unresolved, the visual evidence of the scan endows biological attributions with a sense of facticity and apparent truthfulness. Once a behavior or mood is conceptualized as biological, it enters the discursive province of health and can be legitimately evaluated in terms of health and illness. Again, even if key questions are left unresolved—for instance, the scientific standard of health—simply bringing a mood or behavior into the domain of biology opens it up to assessment in the languages of health. The major message of the rhetorical brain might be amended slightly—from "life is calculable" to "*all of* life is calculable." In part because of the rhetorical force of brain imaging and the discourses it supports, assessments of health and illness can be applied to virtually any aspect of life, no matter how minute or prominent. These assessments transform life into something that can be studied, calibrated, acted on, and evaluated according to various criteria. As I will argue, it is this increasing emphasis on making life manageable that makes health such a powerful governmental discourse. Governmentality depends on individuals' active involvement in their own government, and the languages of health work to frame a number of actions and attitudes as "in one's own best interests." Health is, for the most part, an unquestioned value (who doesn't desire health?), so defining activities in terms of health is a key way of making them appear desirable and worth actively pursuing.

From Structure to Process:
New Visions of Health and Illness

It is not simply the mere ability to visualize the brain that produces significant changes in cultural conceptions of health and illness, but rather the particular visual qualities of these images, or the specific type of brain that is materialized by novel imaging technologies. Two features of contemporary imaging technologies are important: First, imaging technologies are heralded for their powers of spatial resolution, or their abilities to visualize the brain at ever-smaller levels, contributing to what Nikolas Rose describes as the "molecularization" of life.[29] Second, technologies like PET and fMRI are "functional," able to visualize not only the anatomy or structure of the brain but also its processes in time. In popular neuroscience, much of the fervor over recent advances in imaging technologies stems from their heightened powers of temporal resolution, or their ability to visualize the active brain as it is in the actual process of carrying out specific tasks. For example, the neuroscientist Richard Restak describes how images can depict both the structure and function of the brain, but he highlights the latter as the most innovative and exciting: "Thanks to imaging techniques, we can now explore what is actually occurring in the brain as we go about our daily lives."[30] Similarly, *Newsweek*'s "New Ways of Seeing the Brain" celebrates recent functional imaging technologies that "allow researchers to look at the real-time brain activity of living people," a development that has "revolutionized" neuroscience.[31]

Functional technologies capable of visualizing the brain at smaller and smaller levels in "real time" are both producing distinctive images of the brain and propelling the dissemination of different languages for describing the space of the brain. Words like pathways, circuits, and networks, terms that indicate movement, process, and interactivity, are increasingly circulating alongside the older language of regions, structures, and modules, terms that suggest relatively static, bounded spaces. This grammatical shift—born of imaging technologies' much-proclaimed abilities to visualize the active, working brain—is caught up in broader historical controversies regarding brain structure and function. In recent decades, the predominant frame for the study of the brain has been localization.[32] As the psychiatrist Norman Doidge describes it, localization is often explained via machine or information-processing metaphors, and it views the brain as "a group of spe-cialized processing modules," or a machine comprising distinct parts, "each of which performs a specific mental function and exists in a genetically pre-determined or hardwired *location*—hence the name."[33] Localization sciences focus on determining which structures or modules of the brain are respon-sible for which functions. In this view, structure-function combinations are hardwired or biologically determined and, as such, are pretty much set for life in a normally developed adult brain. Thus, because brain development is for the most part fixed, illness or abnormality occurs when a structure is damaged or fails to develop normally.

According to William Uttal, localization is slowly receding from scien-tific favor, and connectionism, an alternative theory of brain function, is growing in popularity.[34] Connectionist vocabularies, as the case studies in this book will reveal, already pervade the ways we talk about the brain in cultural contexts. While localization views specific brain regions or mod-ules as responsible for determinate functions, connectionism tends to view the brain as an array of dynamic processes or distributed neural networks. Doidge writes, "A brain system is made of many neuronal pathways, or neu-rons that are connected to one another and working together."[35] Because these connections are mobile and dynamic, the brain can create new path-ways, as well as strengthen or weaken existing pathways. Connectionism is also sometimes referred to as distribution because proponents believe that neural networks distributed throughout the space of the brain (rather than discrete regions or modules) accomplish brain functions.

Connectionist theories, compared to localization theories, tend to be far more accommodating of the languages of brain plasticity. Because localiza-tion views brain modules as hardwired to perform specialized tasks, damage to a structure is likely to impair the associated function permanently. Con-nectionism, on the other hand, understands the brain as far more malleable. A structure is ultimately just a space or hub of neural connectivity, and if a region of the brain is damaged, new hubs can be wired through various interventions designed to reshape the brain and restore function. A given area, in other words, is not predetermined for any particular function, and different circuits or pathways can modify themselves to perform different tasks. This understanding of malleability shapes conceptions of both cure and optimization: the brain's ability to rewire itself is not limited to cases of injury or brain damage, but is also taken to apply to healthy brains that can be rewired for more optimal function.

Once again, Kawashima's brain training program helps to conceptualize how images of brains as dynamic arrays of processes are changing the ways we talk about brain health. As Kawashima articulates in his book (in language repeated in *Brain Age* marketing discourses), healthy brain process entails both an abundance of neurons and an abundance of strong connections among neurons. Both the number of neurons and the quantity and quality of neural connectivity can be increased, according to Kawashima, through routine exercises. The theme that connectivity is the arbiter of brain health is graphically illustrated by the figure featured in an advertisement for *Brain Age*. The image depicts a man with a shaved head playing *Brain Age*, with an expression of rapt attention on his face. Lighted points crosscut the man's shaved head, with lighted lines connecting the dots to produce a geometrical "net" on his head. The pattern of dots and lines suggests that the man's activity (playing *Brain Age*) is both "lighting up" his brain (a common phrase used to describe the patterns of brain activity revealed by scientific imaging) and producing the highly interconnected grid mapped on his head. The graphic does not depict the structure of the brain, but instead articulates the brain as a series of points (neurons) and connections, a map of interactivity that can be improved through the proliferation of points and strengthened connectivity. The brain, in this rendering, is less of a thing, in the sense of a dense biological organ, and instead appears as something that is nearly virtual—a space of activity and dynamism that can be diagrammed but not captured in visible form as something resembling actual organic material.

Although connectionism is a growing force in scientific and popular contexts, localization theories continue to exert considerable persuasive appeal. I view connectionism and localization from a rhetorical perspective, and I understand them as two vocabularies or patterns of articulation that merge and interact in diverse ways. In other words, I am less interested in whether either theory is true or scientifically validated; rather, I am primarily attentive to how languages of connection and location circulate in popular neuroscience and condition cultural conceptions of both the brain and human identity. Moreover, from a rhetorical perspective, connectionism and localization are not mutually exclusive. In popular neuroscience, there is considerable slippage between languages of localization and languages of distribution and connectivity. For example, the two vocabularies often cohabit discourses that identify distinct regions but view them as interconnected and interdependent. In addition, theories that highlight particular areas of the brain for targeted interventions are frequently articulated in blended languages. In these permutations, regions are responsible for distinct functions, but regions themselves are portrayed as areas of malleable activity that can be intensified, dampened, or balanced through targeted intervention. In addition, localization remains an especially attractive language for understanding the brain in contexts supporting determinist themes. In claims that situate the brain as responsible for emotions, thoughts, or behaviors, it is persuasive to pinpoint a specific region of the brain and hold it responsible for some particular experience. In claims that highlight human agency and

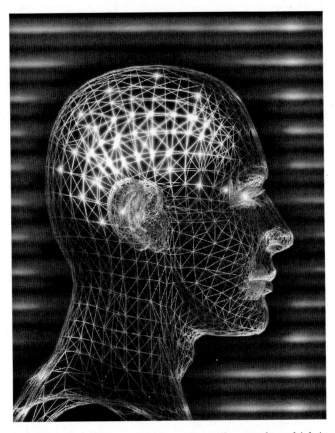

FIGURE 2 This figure, suggestive of neural networks—which is similar to the one featured in Nintendo DS's advertisements for *Brain Age*—appeared on the cover of NOVA's 2007 *Secrets of the Mind*. The PBS DVD featured the neuroscientist V. S. Ramachandran, whose research has contributed to the growing interest in brain plasticity.

Chad Baker/Getty Images

the potential to willfully transform some aspect of life, connectionist languages tend to be more appealing because they more easily accommodate notions of plasticity, change, and dynamic process. In chapter 2, I take up the tensions between localization and connectionism in greater detail, paying particular attention to the legacy of phrenology (perhaps the quintessential localization theory) in contemporary brain science. The continuing oscillations and permutations of connectionist- and localization-oriented theories suggest that popular neuroscience is not monolithic, but rather includes competing and sometimes contradictory themes.

Although localization continues to be a foundational discourse of popular neuroscience, the proliferation of languages of process, connection and plasticity are contributing to significant transformations in scientific and cultural conceptions of health and illness. In the next section, I take a more detailed look at how functional brain images are enabling articulations of health as capital, or a resource that can be maximized, accumulated, and calculated. As I will explain, the articulation of health as capital is an important contributor to broader changes in social organization that Gilles Deleuze has described as a shift to a "control society," marked by a general fragmentation of institutions in tandem with a growing emphasis on individual responsibility for managing risks and optimizing productivity. Popular neuroscience directly enables these broader shifts by situating individuals as continuously responsible for pursuing and optimizing their own health, a comprehensive project that encompasses almost the entirety of life—due in part to the expansionist definitions of health supported by brain imaging research.

Health, Control, and Capital

It is interesting that for Bush's council, the problem of how to safeguard an "authentic" human existence is essentially a problem about language, or the terms through which we frame and interpret our lives: "We will need to hold fast to an account of human being, seen not in material or mechanistic or medical terms but in psychic and moral and spiritual ones."[36] According to the council, it is only by holding fast to a humanistic vocabulary that we can enjoy the fruits of biotechnology without succumbing to its dangerous temptations. Despite the warnings of Bush's council, humanistic vocabularies are being supplanted, or at least heavily supplemented, by therapeutic vocabularies that medicalize a host of human experiences and behaviors that were formerly understood in spiritual or moral terms. Contemporary neuroscience, as Steven Johnson puts it in *Mind Wide Open: The Neuroscience of Everyday Life*, "presents us with a new grammar for understanding our minds."[37] Common ways of speaking about the self and others are increasingly suffused with terms from neuroscience, such as "serotonin," "chemical imbalance," and "neurotransmitters." In today's "antidepressant era," brain chemistry and the details of refinements in designer compounds like Prozac are "the stuff of fashionable coffee table talk."[38]

This rhetorical shift characterized by a diffusion of biological—and specifically neuropsychiatric—terminologies throughout social discourse is part of larger transformations in the conceptions and practices of health. In contemporary neoliberal society,[39] health has become a type of capital, or a resource or value that can be accumulated, stored up, and channeled for individual and social productivity.[40] Put simply, health is not a default

state of existence, but rather something to be achieved in ever-greater incre-
ments. It is like a resource that can never be maximized or fully realized—one
can always have more health or achieve a greater state of health. Along with
the notion of brain age, Kawashima's concept of "neural reserve," sometimes
called "brain reserve" in other media, illustrates the ways in which health
functions as capital. The idea of neural reserve implies that through calculated improvement efforts, individuals can accumulate increasing quantities
of health; further, this can essentially be stored as "reserves" that one can
draw from to compensate for brain decline brought about by natural aging
processes, or to "cushion" the effects of exposure to risky stimuli. Reserve is
something that can be assigned a quantitative value: neuroscientist Daniel
Amen writes, "The more reserve you have, the healthier you are. The less
reserve, the more vulnerable you are."[41] In popular neuroscience, brain
reserve is often linked to connectionist understandings of the brain. Richard
Restak, for example, describes brain reserve in terms of the brain's capacity
for connectivity, explaining that building up the strength of neural networks
can compensate for future decline in the number of neurons. Individuals are
encouraged to start a brain fitness program early in life to build up the greatest possible store of brain reserves and "offset" the natural effects of aging.[42]
If the brain consists of many strong connections, then the loss of individual
neurons will have minimal effect on overall brain function.

When health is conceived as something that can be accumulated,
quantified, and leveraged for future gain, it is unmoored from its binary
position opposite illness, and takes on the characteristics of a free-floating,
generalized discourse of control. I am using control in a technical sense
here, referring to the ways in which the value of health has come to frame
multiple aspects of individual and social life as calculable and manageable
loci of intervention.[43] Control is Deleuze's term for a society marked by the
spread of market models into different arenas of human life. For individuals, the pursuit of health becomes like the pursuit of money; it is a resource
to be maximized through a series of cost-benefit calculations applied to the
minutiae of daily living. At the level of society, health is an efficient means
of controlling or governing the population by encouraging self-supervision
and reducing the need for top-down institutional regulation.

Control and the Intensifications of Discipline

A brief detour highlighting the distinctions of control as a framework
for understanding social power helps to illuminate the cultural force of
popular neuroscience and brain imaging. Discipline and control, in addition to other theoretical terms I will introduce in this chapter, including
"normation" and "normalization," are useful shorthand for conceptualizing
popular neuroscience's role in facilitating social change. Deleuze, drawing

on Foucault's discussions of security, introduced the notion of control to describe the intensifications of disciplinary power that mark contemporary culture. Foucault's theory of disciplinary power has become a familiar and useful vocabulary for talking about power, primarily because it views power in terms of participatory, constitutive relationships rather than something that is imposed from the top down. Disciplinary power is not solely a repressive power that stymies self-will; rather, disciplinary power works at the micro-levels of society to shape individuals' desires and sense of self to produce subjects who actively work on themselves to guarantee their social productivity and fitness for social roles. Disciplinary technologies, or ways that individuals can know and act on their selves, include surveillance, the exercise of "infinitesimal power" over the body (including supervision of its movements, gestures, and attitudes), and the codification and partitioning of time and space (e.g., scheduling regimens).[44] In Foucault's theorization, disciplinary power was primarily activated and deployed within the spaces of institutions (including prisons, factories, and schools), with the primary objective of achieving conformity to a norm or standard of perfection. This type of conformity occurs, as Michael Hardt and Antonio Negri explain, not through top-down impositions of power, but rather through processes that make conformity appear both voluntary and essential.[45]

In recent years, scholars have increasingly focused on the ways in which disciplinary power is extended and intensified in a society marked by the fragmentation of institutional spaces and the construction of individuals as entrepreneurs of themselves, responsible for managing risks to their well-being while optimizing their contributions to society. Mechanisms of disciplinary power are detached from institutional contexts and become "free-floating," exercising power over life in a spatially and temporally continuous fashion, and not according to the stop-and-start rhythms created by individuals' movement from institution to institution.[46] These extensions and intensifications constitute, in Jeffrey Nealon's terms, a mutation of disciplinary power.[47] This mutation, or what Deleuze describes as the emergence of the control society, can be summarized into two general themes or principles. First, discipline becomes more and more interiorized within individuals' bodies and, more important, their minds or psyches. In other words, disciplinary technologies are more mobile and immanent—they are, in a sense, always with individuals, carried around within their own subjectivities, and continuously heard as "incessant whisperings" that come from within.[48] Second, the process of ensuring conformity to a norm mutates into a process of self-optimization. Because health—and mental health in particular—has functioned as a predominant discourse of discipline and conformity, I return to examples from popular neuroscience to trace the ways in which brain images are working to enable and solidify this mutation of discipline to control.

New lenses for understanding the brain supported by functional imaging technologies are fundamentally changing conceptions of brain health. In part because of the way images depict the brain, scales and terminologies for assessing the activity rather than the structure of the brain are becoming widespread. Instead of binary evaluations (the brain is either healthy or ill), brain activity is assessed in languages suggesting a graduated scale or spectrum model of evaluation, with terms including more or less, slower or faster, and stronger or weaker. These languages resonate with the types of descriptors applied to brain scans themselves. Scans, which typically depict brightly colored "hot spots" that fade in intensity, are often evaluated in graduated and comparative terms, including "hotter," "brighter," "darker," and "more active." Terminologies of scale and spectrum allow for comparisons between different brains and scans, but they do not offer a clear-cut benchmark or definitive "norm" that serves as a universal ground for measurements. As I will explain in more detail, the diversity and dynamism of brains revealed by imaging technologies make such a standard or definitive norm untenable.

Along with changing assessment vocabularies, the goal of interventions shifts from repairing a deficient structure to tweaking a dynamic process. For example, if an identifiable brain structure is judged to be sick or abnormal, generally the goal of treatment is to correct for the deficiency and restore a predetermined standard of health. When the focus of assessment shifts to active brain processes, however, intervention comes to include a whole range of often-subtle practices of calibration, balancing, and optimization. For example, brain training manuals encourage readers to pursue "better" brains by intervening at the level of process to balance, slow down, speed up, intensify, or dampen activity levels. Similarly, ads for antidepressants and other psychotropic medications promise to "stabilize" or "balance" chemicals as they course through the circuits of the brain. Functional brain images and their accompanying languages suggest that brain activity, and hence life, is precarious—its continuity depends on an optimal balance and pace of activity, and it is constantly threatened by disruptive influences. Thus, having a healthy brain is not as simple as preserving a structure or achieving and maintaining a normal state; rather, it requires a constant, ongoing attention to and calibration of subtle, dynamic processes at incredible levels of detail.

Normation and Normalization

The distinctions between discipline and control are significant insofar as they offer a vocabulary for talking about power that helps to understand what is at stake in brain culture. Foucault's concepts of normation and normalization help to both clarify the sometimes-subtle distinctions between discipline and control, and also to illuminate how popular neuroscience's growing emphasis on active brain processes enables recent mutations in disciplinary power. "Normation" is a type of judgment that generally corresponds to disciplinary power, while "normalization" is associated with control. In the context of discipline, power functions through the imposition of a norm or standard, a predetermined "point of perfection."[49] Disciplinary power first posits the norm and then "consists in trying to get people, movements, and actions to conform to this model."[50] Individuals adopt and internalize this norm, and then act on themselves in their attempts to live up to this standard. Although individuals act to achieve these norms out of their own apparent volition, the norms originate externally, in social institutions or cultural expectations—these norms have, in Foucault's terms, "an originally prescriptive character."[51] Individuals are not subjected to a single norm throughout the course of their lives, but rather multiple, often contradictory norms. These norms vary according to different institutional spaces or social contexts. For example, in the workplace, individuals strive to be perfect workers, and in the home they try to become perfect parents. Deleuze notes that this spatialization of disciplinary power creates a stop-and-start rhythm of sorts: "In disciplinary societies you were always starting over again (as you went from school to barracks, from barracks to factory)."[52]

Normalization, on the other hand, encourages the management of natural processes in order to maximize gain and minimize harm, incorporating a sense of dynamism and open-endedness into its schemes of assessment and intervention. Normalization is still normative, in the sense that it incorporates an element of "should," as in "you should be better," or "you should do this," but this "should" does not operate in relation to a predetermined, fixed standard of perfection.[53] A simple way of thinking about this shift from normation to normalization might be as a change from "you must be like *this*," where "this" is a particular norm or model, to "be all you can be!" In the latter expression, the ceiling for improvement is completely removed—you no longer have to work to make yourself compliant with a pre-given standard, now you are enjoined to continuously work on yourself to be better, without limit or end. You can never stop being better, because in control, there is no "perfect model" to finally achieve.

Popular neuroscience is deeply implicated in this mutation from normation to normalization. Because popular neuroscience speaks the languages of optimization and improvement, it encourages continuous work on the self, unconstrained by any definitive, predetermined standard of health. Moreover, brain culture changes the stop-and-start pattern of disciplinary power, or at least modifies it in key ways. Brain images are used to support claims that being a good teacher, student, or parent depends on having a healthy, optimally functioning brain. Working to optimize brain function is the path to improvement regardless of where you are (school, home, work) or what area of life you are particularly concerned with (studying, parenting, working). The brain becomes, in essence, the site or space for all optimization energies, independent of the specific outcome or type of success that is desired. In terms of the shift from normation to normalization, this has two consequences. First, control becomes continuous instead of punctuated by start-and-stop rhythms. Regardless of the particular institutional space one inhabits, the basic target of self-management is the same: the optimization of brain function. For example, as Elkhonon Goldberg argues, the brain does not have "thematically specialized modules" or "special purpose neural circuitry" responsible for distinctive tasks or norms. The "basic mechanisms" of brain function are the same across "economic, moral, or legal contexts," and thus calculated attempts to optimize these basic mechanisms by enhancing connectivity are the appropriate means for achieving success in any area of life.[54] Restak similarly instructs readers, "You must engage in repetitive exercises that set up the relevant circuits and sharpen their expression. *This holds true whatever your goal and whatever degree of mastery you seek.*"[55] Thus, not only is control through brain optimization continuous across different social spaces and areas of life, it is almost entirely interiorized. Individuals literally carry around within them the "space" on which they are to act to produce themselves as better (again, whether one desires to be a better student, teacher, or worker is of little consequence, as the basic technologies of brain health optimization are quite consistent across these areas).

Second, popular neuroscience not only makes control spatially continuous by channeling optimization energies to brain function, it also makes control temporally continuous by removing any endpoint or final stopping point in the pursuit of better brain health. In other words, optimizing brain health is temporally continuous in the sense that it is never completed, as there is no predefined norm to achieve. Unlike normation's dependence on a fixed goal of perfection, control's goal is contingent, both because it relies on processes of maximizing potential and minimizing risk, and because it addresses highly individualized entities. Initially, control does not have a set standard, but rather works to maximize and minimize the respective potentials and risks of a particular brain (and hence individual life) through calculated interventions. In the context of health, these interventions take the form of technologies of enhancement and prevention. "Enhancement technologies," vilified by Bush's council, promise to make individuals, in Kramer's pithy phrase, "better than well."[56] Examples of enhancement technologies include mood altering drugs such as Prozac and Ritalin, which improve the functioning of individuals even if they are not depressed, and drugs such as Viagra, which improve sexual performance even in the absence of definitive pathology. Preventive health technologies are the flip side of the "cosmetic pharmacology" coin. Like enhancement technologies, illness- and disease-prevention discourses situate health as something that must be calculated, managed, and channeled for maximum profitability. The contemporary emphasis on prevention, Paul Rabinow writes, is above all "the tracking down of risks," a mode of surveillance that attends to "likely occurrences of diseases, anomalies, deviant behavior to be minimized, and healthy behavior to be maximized."[57] Individuals can be "asymptomatically or pre-symptomatically ill" and require "treatments" far in advance of any visible pathology.[58] Enhancement and prevention are related modes of submitting life to a logic of accumulation; through the twin strategies of optimization and minimization, one can capitalize on one's current allotment of health.

In addition to its twin emphases on risk management and optimization, normalization addresses brains that are understood as unique. An individual brain has various strengths and weaknesses, and the goal is not to flatten out these "peaks and valleys" to conform to a universal model, but to work with each unique landscape to capitalize on its strengths and minimize the risks posed by its weaknesses.[59] Processes of maximization and minimization must be individualized, because imaging shows that no two brains function in exactly the same fashion. It is not so much that some brains are healthy and others are sick: rather, because all brains are different, they all require different types of interventions to capitalize on their unique potentials for health—in essence, the norm is entirely personalized. For instance, Restak explains that brain scans suggest "individual 'signatures': patterns of brain activity that vary from person to person."[60] Each person must engage in brain health activities that capitalize on their unique brain signatures or "styles" of brain function. Individual differences in brain activity, Goldberg writes, can be viewed as "*multiple expressions of normality*," and when faced with two distinctive styles of brain function, "we assume that both are normal, but different."[61] Recently, the variability of individual brains as revealed by scanning technologies has led to celebrations of "neurodiversity" and activism to gain recognition for diverse brain configurations.[62] Thus, with normalization, diversity is a resource rather than a barrier to conformity. This capitalization of brain differences further interiorizes the obligations to improve one's self, because it is a matter of "becoming your own best self," which is far more personal than becoming "the best worker you can be" or "the best student you can be." The motivation for optimizing activities, then, is not a norm generated externally before being interiorized, but a desire that must originally emanate from within the depths of the psyche. In Rose's words, "Disciplinary techniques and moralizing injunctions as to health, hygiene, and civility are no longer required; the project of responsible citizenship has been fused with individuals' projects for themselves. What began as a social norm here ends as personal desire."[63]

When health saturates the routines of daily existence as a personal desire, regulatory responsibility moves outside the province of institutions and becomes, at least in part, an obligation of individuals. The deinstitutionalization of health—mental health in particular—can be witnessed in a variety of indicators, including the terminological shift from "patient" to "consumer"; the growth of health-seeking information on the Internet; and the prevalence of self-diagnosis, largely fueled by the explosion of direct-to-consumer pharmaceutical advertising, whereby citizens are coming to doctors requesting specific diagnoses and medications. As I will argue in more depth in later chapters, these changes are wrapped up in languages of freedom, choice, and respect for individual diversity, but despite alluring promises, these discourses carry real consequences as they entwine individuals in intensive obligations framed as innermost desires and make health at once more compulsory and more difficult to achieve.

Rhetorical Formations

A foundational premise of this book is that the social functions of health and illness are rhetorical. It is noteworthy that the council's disapproval of biotechnological therapies replicates traditional charges against rhetoric. The denunciation of biological techniques that go beyond therapy suggests that biotechnology should be begrudgingly tolerated so long as it remains within its appropriate boundaries and does not attempt to move beyond a curative or restorative function. When it goes past its circumscribed territory, biotechnology corrupts the soul, rendering the subject an inauthentic imitation of true human essence. Rhetoric, as the "handmaiden of philosophy," has been similarly limited by dictates that it remain subservient to "Truth." When rhetoric roams beyond its territory, it, too, becomes imitation and falsehood, the antithesis of everything that is real and authentic.

The difficulty in both cases is that "authentic human essence" and "Truth" are always already inhabited by their Other, technologies and rhetoric, respectively. Contemporary neuroscience recognizes that everything from conversations and attitudes to chemicals and surgical interventions change brain function. It is impossible to neatly carve out a "natural" state that unproblematically bears the label "authentic." Similarly, whatever the status of "Truth," our encounter with it is always mediated, whether by languages, images, or other rhetorical forms. Rhetoric saturates our society and our selves, and it is impossible to bracket it off from the rest of existence. Similarly, it is impossible to mark off the "natural self" from the malleable self that is changed through interventions. We are, in short, biological *and* rhetorical creatures, and attempts to distill a human essence that strain out either or both of these constitutive forces is bound to fail.

In making this claim, I do not want to suggest some ontological chasm that separates rhetoric from material biology. One of my theoretical commitments is to the premise that rhetoric is material: it is a social force that has real, tangible effects. In this book, I trace this material force by examining various manifestations of the rhetorical brain as it circulates in diverse contexts—self-help books, a science museum, popular magazines, and public policy discourse. I engage these texts as part of what Celeste Condit describes as a rhetorical formation.[64] Rhetorical formations are historically specific patterns of discourse, or recurring rhetorical forms—images, metaphors, narrative devices. Tracing rhetorical formations means attending to the interactions and the variances in public discourse as it flows throughout diverse social arenas. Specifically, a rhetorical formation is defined by three characteristics: it is multiple, contested, and dynamic. First, rhetorical formations are multiple because they are not defined by a single element or type of element, but by combinations and flows of different rhetorical trajectories. The model of rhetorical formations is premised on the fact that texts cannot be neatly isolated and divorced from their contexts. Instead, rhetorical texts are understood as "focal nodes in a larger torrent of human discourse."[65] One consequence is that the "method" of engaging rhetorical formations is inevitably messy—it is difficult to draw neat lines around a precise object of study. In this sense, rhetorical formations can also be understood as assemblages, or collections of diverse elements that somehow get linked together or connected. The vocabulary of assemblage highlights the contingency of rhetorical formations. A formation, in this sense, is not determined or static, but rather emerges from the contingent interactions of diverse rhetorical entities. In this book, I look at a wide range of elements that operate in seemingly very different areas of life—public policy, entertainment, education, family life, and so on—but are connected, in sometimes surprising ways, through a shared attention to the brain.[66] Second, even when particular rhetorical formations are dominant they are never determinant; they are always open to contestation and negotiation. Finally, rhetorical formations are dynamic—they comprise constantly shifting patterns, so they are difficult to pin down with neat definitions for any length of time.

This model is extremely valuable for my purposes, because I view popular neuroscience discourses as nodal points in larger discursive flows associated with biological materialism and healthism. There is no single text that instantiates this configuration; rather, it is the product of an array of discourses that circulate throughout an array of various conversations. I engage a rhetorical formation, or zone of discourse, frequented by neuroscience, psychiatry, and biomedical imaging of the human brain. My area of focus inhabits a region described by Rose as "psy," populated by the psycho-sciences, including psychiatry, psychology, and related disciplines.[67] Psy is neither monolithic nor static, though Rose identifies a series of "family resemblances" among the discourses that traverse its terrain. My focus is best conceived as a province of psy, habitually traveled by those discourses most closely related to contemporary neuroscience with its emphasis on brain biology and its dependence on novel brain imaging technologies. Within this discursive zone, I am generally oriented toward the popular or public dimensions of these discourses rather than the types of communication and persuasion internal to scientific and medical practice. This book explores a wide array of texts, from popular magazine articles to documentary films, from nonprofit websites to popular fiction to political speeches. My analysis incorporates both the visual and verbal features of these texts, including graphs, pictures, charts, and language. Each chapter focuses on a very specific text or network of texts but engages themes that persist throughout the broader rhetorical formation. I see each of these case studies as a way to assess the function of the broader rhetorical formation while attending to the particular discursive features of specific texts.

Additionally, the perspective of rhetorical formations leads me to focus on patterns of discourses—recurrent images, vocabularies, and techniques of arrangement—across multiple texts. I am less concerned about the specific authors of these texts, because these patterns suggest broader sociocultural trends that are in a sense bigger than any single author or institution. My approach to psychiatric culture differs from criticisms that focus on the power of psychiatric pharmaceutical corporations and their economic influence. My approach to psychiatric culture is certainly critical, but it is not a "dominant ideology" approach that views psychiatric power as something imposed by greedy corporations. This latter approach is wanting for two reasons: First, pharmaceutical companies do play a notable role in these events, but they are not alone. A number of different actors—individuals, families, schools, museums, government agencies, research scientists—are involved in the emergence and transformation of psychiatric culture, and social effects can rarely be traced back to a single actor. Second, the success of these pharmaceutical companies is in part conditioned by changes in the public grammar. Brain rhetorics are highly persuasive, and they are partially constructed through the active participation of public audiences. In other words, pharmaceutical success is both partial cause and partial effect of the neuroscience revolution, and to single corporations out for blame is to ignore the complexity of these issues.

That said, I am critical of brain culture and the ways in which popular neuroscience frames the self and society. Popular neuroscience speaks in the heady languages of personal freedom and social liberation, promising that through knowledge and control of the brain, we can create better selves, better communities, and a better world. I am deeply suspicious of this teleological impulse that has inhabited strains of scientific discourse for centuries. The type of freedom popular neuroscience heralds is ultimately a highly constraining freedom, or a version of freedom that comes at considerable cost. It ties individuals to projects of self-improvement without end, situating them as ultimately responsible for personal and collective failures to achieve perfection in physical, social, and, above all, economic contexts. The self-entrepreneurial modes of living disseminated by popular brain rhetorics oblige individuals to be free, in Rose's words, entwining freedom with relentless obligations of self-scrutiny and self-management. These personal projects of self-perfection feed into neoliberalism's broader agendas of personal responsibilization and market models for social services. In the final chapter, I examine some of popular neuroscience's teleological claims and situate them in the context of recent changes in social organization, providing critical commentary on both the promises and dangers of brain-based ways of understanding human life.

Chapter Preview

To say that the brain is rhetorical is to suggest that it is not always the same—that it is contingent on numerous social, political, and scientific factors. In chapter 2, I outline the features of the "New Brain," or the contemporary rhetorical brain, using Restak's recent book *The New Brain: How the Modern Age Is Rewiring Your Mind* as a touchstone for a brief history and overview of contemporary neuroscience and brain imaging.[68]

After the historical and theoretical overview, I offer four chapters, each devoted to a specific text or configuration of closely related texts that function within our psychiatric culture. These case studies generally move through different tiers or concentric circles, starting with the constitution of individual identity and moving to government policy formation, although these issues intersect in ways that make this type of clean demarcation impossible. Because a key premise of this book is that the rhetorical brain coordinates individuals, social units, and government agencies, these circles are overlapping, and the case studies point to many common features.

In chapter 3, the first case study, I examine the circulation of brain images in self-help books. Describing a genre I refer to as "brain-based self-help," I focus on Amen's recent publication *Making a Good Brain Great*. Amen, a well-known brain scientist and clinical psychiatrist, is recognized for using imaging technologies for clinical purposes.

In chapter 4, I examine popular discussions of baby brain health, as they have appeared in representative issues of *Time* and *Newsweek* since 1990. Over the past two decades, the baby's brain has been thematized through imaging technologies as a space of attention demanding intervention on the part of social actors, including parents and caretakers as well as educators and physicians.

Chapter 5 addresses the ways in which psychiatric languages are distributed to individuals as terms for expressing personal, subjective experiences. I focus on one key nodal point for the distribution of such languages: *Brain: The World Inside Your Head*, a museum exhibit that opened at the Smithsonian in 2001 and travels to science centers and museums throughout the country.

In chapter 6, the final case study, I trace the rhetoric of recovery throughout recent policy discourse, focusing on George W. Bush's New Freedom Commission on Mental Health Care, a federal agency given the responsibility of comprehensively reviewing the state of mental health care in the United States. The commission's landmark report, *Achieving the Promise: Transforming Mental Health Care in America*, illustrates how popular conceptions of the brain, shaped by neuroscientific words and images, condition social and political structures at the highest levels of government. Although I focus on Bush administration discourse, the rhetoric of recovery is relatively continuous across all the presidential administrations since that of Jimmy Carter.

This book surveys the rhetorical forces at play in contemporary psychiatric culture, moving full circle from an examination of the ways in which individuals come to take up neuroscientific vocabularies to frame their experiences and desires, to mapping the consequences of these changes in the public lexicon as they ripple across the sociopolitical landscape. In short, this is not a book that is just about the brain, nor is it just about science and scientists. It is a book about our selves, our culture, and our historical context. The brain is simply a convenient topos, or a place for speculating about who we are and what we might be becoming. The brain has often played this role, serving as a central node for reflection about the broader characteristics of a historical moment. Throughout history, a culture's understanding of and visualization of the brain can be seen to resonate with other contemporaneous factors—economic, social, political, and ideological. In other words, it is possible to discern often-complex relationships between the ways a people thinks about and sees the brain. The next chapter takes a closer look at these relationships, which I refer to as resonances, to highlight the singularity of the rhetorical brain and to assess more carefully its significance in contemporary American culture.

2

Visualizing the New Brain

The brain visualized by digital imaging technologies has unique properties, quite distinctive from those of the wet brain. Popular neuroscience often differentiates the digital brain and the wet brain by their particular relationships to scientific knowledge. The wet brain in its "natural state" willfully resists the advances of science and openly "defies our attempts to come to grips with it."[1] As the neuroscientist Richard Restak describes, the resistant wet brain is the Old Brain, "remote and mysterious, deeply hidden within the skull and inaccessible except to specialists daring enough to pierce its three protective layers." Due to the "risks involved in plumbing its depths," brain scientists knew little about the brain, and they "certainly searched in vain" for answers to the most fundamental questions about human nature and the relation of the brain to personality, emotions, and behavior. Restak contrasts the Old Brain to the New Brain, the brain that does not require "dangerous intrusions," but rather is easily accessed by imaging technologies. These technologies "reveal exquisitely subtle operational details and provide windows through which neuroscientists (brain scientists) can view different aspects of brain functioning without opening the skull or performing risky procedures."[2]

Although Restak's enthusiasm about science's ability to access the brain and render its inner workings transparent to the scientific gaze relies on a distinction between the accessible New Brain and the defiant Old Brain, his excitement echoes earlier proclamations celebrating science's abilities to "lay bare" the Old Brain. Claims of impending scientific omniscience—ideas that in a very short period of time, science will fully unlock the keys to human identity—are a constant throughout the history of brain science as each new scientific technology and discovery promises to totally reveal the mysteries of nature. For instance, Burford Rawlings, an administrator for the

National Hospital for the Paralysed and Epileptic, wrote in 1913, "The structure and working of the brain had been laid bare, and the stupendous fact
had been established that to each of the cerebral hemispheres were allotted
functions distinct and separate." Brain scientists, pursuing their conquest
with "untiring patience," "demonstrated that every individual portion of the
seemingly homogeneous organ was allotted its own particular task, and in
response to the probing interrogations of science every fibre and filament of
the complex structure yielded up the secret of its being."[3]

These two descriptions, separated by ninety years, participate in a trope
that has inhabited scientific discourse in various forms since at least the
days of Bacon—the construction of nature as a feminine agent that gives
up her secrets to science, in response to either seduction or force. Richard
Restak (whose recent book is titled *The Naked Brain*) describes the New Brain
as an entity that gives itself over to scientific knowledge without the necessity of "dangerous intrusions" or "risky procedures." Contemporary brain
scan technologies are framed as noninvasive "windows," rendering the brain
transparent as it willingly offers up its secrets to the scientific gaze. Here, the
brain is the active discloser of mysteries, unveiling itself before the passive
eyes of scientific technologies. By contrast, in Rawlings's much earlier narrative, the brain is figured as an object of conquest that is won only through
forceful violation. The brain is "laid bare" as it submits to "the probing
interrogations of science," and it "yields up" its secrets only in concession to
the "untiring patience" of determined scientific rigor.

These two descriptions can be taken as representative anecdotes sum-
marizing general social and scientific understandings of the New Brain and
the Old Brain, respectively. As I will explain, the New Brain, or the brain
that is visualized by contemporary imaging technologies and increasingly
understood in terms of distributed connections, is not an entirely new
understanding of the brain, but rather is the most recent stage in the ongoing reinvention of the Old Brain. There are, in other words, many points of
continuity between the Old and New Brains, as conceived in both popular
and scientific contexts. Moreover, there is no single, uniformly accepted
theory of the brain that guides contemporary neuroscience, and most current theories are somewhere in between Old Brain (localization) and New
Brain (connectionist or distributed) paradigms. Instead of conceptualizing the Old and New Brains as two distinctive theories that govern clearly
demarcated historical periods, the Old and New Brains are better understood
as two explanatory tendencies that have framed investigations of the brain
for centuries, with one or the other tendency gaining in relative popularity
in different social and scientific contexts.

One important point of continuity that persists across the diverse conceptions of the brain is a firm commitment to a materialist perspective that

all of what makes us human—our personality, behavior, attitudes, emotions, relationships, and thought processes—can be explained by the brain and accurately accounted for in the languages of neurobiology. A second, related point of continuity is that because both the Old and New Brain perspectives hold that the biological brain contains the answers to our most fundamental questions about human being, proponents of both views are optimistic that scientific investigations of human biology (whether framed as seduction or violent conquest) will result in complete and total knowledge of the brain and, ultimately, total comprehension of human nature.

Despite these similarities, there are profound differences. The most significant difference is the relationship between the structure and the function of each brain. The Old Brain is structured into distinct regions, and each of these clearly divided spaces, or modules, corresponds to a specific function or operation. Rawlings describes the Old Brain as something that has distinct boundaries, and each "individual portion" of the brain is "allotted" its "own" task. The tasks or functions are "distinct and separate," "particular tasks" that can be isolated. For each separate function, there is a distinct structure—thus a one-to-one relation of strict correspondence characterizes function and structure in the Old Brain. I refer to this localization model of brain mapping as analogical because there is a relation of identity posited between two elements, the structure and function, such that each function corresponds to and is represented by a single structure. For example, in the early 1800s, phrenologists posited a one-to-one relation of representation between brain structures (visible by way of analysis of the skull) and mental functions or "faculties," such as conscientiousness, hope, and combativeness. Each bump, nook, and cranny of the skull represented or corresponded to a single identifiable function. Restak's New Brain presents a different sense of relation between structure and function. The New Brain does not have the clearly identifiable boundaries or precise divisions of the Old Brain; it is a region of "subtle operational details," an active "functioning" process. Brain scan technologies visualize "networks and circuitry," dynamic webs and "wiring diagrams" that cannot be reduced to one-to-one matching of structure and function.[4] The New Brain is seen as a field of combinatorial interactions, too dynamic and dispersed to be bound by relations of "correspondence." In the transformation from the Old Brain to the New Brain, something like a digital logic of combination comes to replace the analogical pairing of structure and function. With the New Brain, functions are envisioned as the effects of both local and long-distance interactions between mobile elements that traverse the space of the brain, effects often described as electrical and chemical forms of communication. In the case of the New Brain, the brain is still the causal agent that produces mental functions—the difference is that these functions are not produced by a single structure, but rather by dynamic and flexible connections of distributed structures, or, more accurately, distributed networks comprising minute elements such as neurons and molecules that are conceptualized on a much smaller scale than the structures of the Old Brain.

Locations and Connections: What Is at Stake?

The ongoing controversy between localization theories and connectionist theories is not an arcane scientific dispute with little relevance for everyday life. A central premise of *Brain Culture* is that the ways we visualize and talk about the brain are intimately caught up with our culture, including both how we practice our lives as individuals and how we organize our social worlds. I view localization and connectionism as two discourses or grammars comprising both vocabularies and visual images that work together in various ways to shape the rhetorical brain. In this chapter, I trace the interactions of the discourses of connection and localization through three different articulations of the brain in order to answer two questions. First, how do images and languages of the brain operate in conjunction with broader cultural trajectories to condition particular understandings of human nature and social organization? In other words, what are the diverse, mutually constitutive relationships between understandings of the brain, cultural practices, and conceptions of human identity? And second, why is it that the brain enjoys a privileged status in cultural life? In other words, why has the brain, more than any other organ, come to be revered as a special, even sacred entity endowed with unique powers? While the brain's centrality might be explained with recourse to scientific fact or historical data, I answer this question from a rhetorical perspective, arguing that historically contingent patterns of talking about and seeing the brain shape its status as an exceptional organ blessed with extraordinary abilities.

I begin this chapter with a brief theoretical interlude to make clear my own commitments in relation to science, truth, and language. This section will give readers a better sense of the perspective I bring to the brain rhetorics examined throughout *Brain Culture*. After this section, I offer a historical-rhetorical account of popular neuroscience. I sketch a basic history of localization efforts that have driven brain science for centuries. Against the backdrop of this historical sketch, I compare and contrast three different versions of the brain as it is shaped in cultural discourses: the Old Brain of phrenology, with its strong localization tendencies; the intermediate brain of António Damásio as articulated in *Descartes' Error*; and the plastic brain described by Sharon Begley in *Train Your Mind, Change Your Brain*, also published as *The Plastic Mind*. By taking each brain in turn and setting each in historical perspective, this chapter provides a more robust understanding of the complexities of popular neuroscience and a richer conception of what is at stake in brain culture.

Theoretical Interlude

The continuities and discontinuities between the Old Brain and the New Brain illustrate two key premises of this book. First, they show that the history of scientific investigations into the brain is discontinuous. There are resonances that echo across different junctures in the history of brain science, such as the persistently optimistic predictions of total knowledge and scientific conquest, but despite the recurrence of certain tropes and themes, this history resists a transcendent telos or principle of movement. In other words, the history of brain science is not a single, coherent story in which scientists accumulate greater and greater knowledge as errors are progressively eliminated. Instead, if the history could be sketched visually, it would not resemble a single ascending line, but rather a messy collage of multiple intersecting story lines. There are two major categories of discontinuity. First, what counts as science—in other words, the acceptable methods for uncovering the truth about the brain—has changed considerably, and second, the scientific object, or what scientists study as "the brain," has also changed. For example, scientists today study the brain by looking at digital data associated with magnetic and electric currents of brain function. Previously, scientists might have looked at drawings, or the brains of cadavers, or even animal brains. It is rarely the brain itself that is the centerpiece of attention—the actual objects that scientists attend to, and the technologies that produce these objects, vary considerably across history.[5] Michel Foucault, describing the history of medical science, puts this eloquently: "The historical tie which the different moments of science can have with one another necessarily has this form of discontinuity constituted by the alterings, reshapings, elucidations of new foundations, changes in scale, the transition to a new kind of object. Error is not eliminated by the muffled force of a truth which gradually emerges from the shadow but by the formation of a new way of 'speaking true.'"[6] Foucault suggests that visual and conceptual apparatuses of knowledge do not possess a universal legibility. The production and consumption of scientific knowledge are regulated by multiple factors, including extra-scientific social and cultural conditions. This rhetorical and epistemological discontinuity is part and parcel of an ontological discontinuity—what the brain *is* in scientific investigation is transformed as it is constituted through different technologies, including concepts, modes of visualization, and laboratory practices. Different visualization techniques, for example, make legible different brains—the wet brain of the autopsy table is not identical to the digital brain of contemporary fMRI scans.

This approach to brain culture is heavily informed by what is commonly called articulation theory.[7] The most important contribution of articulation theory, as advocated by rhetorical scholars and other critical theorists, is that it views truth as something that is real yet dependent on the languages, activities, cultural practices, and social institutions we use to organize our lives. For example, a particular claim to truth—such as "the mind is an effect of the brain"—is real, in the sense that it has force in our lives. This understanding of the mind and the brain leads us to act in particular ways—for instance, when we suffer mental distress, we intervene on the brain with psychotropic medications or cognitive therapies. It is, at the same time, contingent because it depends on a whole series of historically specific vocabularies, technologies, and practices that make this statement legible, or give it a ring of truth. Predominant scientific and cultural understanding of what constitutes mind, what constitutes brain, and what type of relation the two have has changed over time and continues to change in fairly significant ways. Throughout my own analysis of brain culture, I tend to bracket the question of the truth of various scientific and cultural claims about the brain and instead focus on the effects or consequences of these different claims. For example, when tracing the oscillations between localization and connectionism in popular neuroscience, I am less interested in which theory is more valid from a scientific standpoint or more true in some universal sense, and instead I am attentive to the cultural consequences associated with different grammars for conceptualizing the brain. Again, how we understand the brain and the various truth claims associated with it is shaped by predominant vocabularies, technologies, and images, and our truthful discourses tend to undergo modification when new vocabularies, technologies, and images come into play.

The second major premise is related to the first: that different configurations of the brain have a certain relationship to their sociopolitical and economic context, which I describe with the term "resonance." Not only is the history of brain science marked by discontinuity as the object that scientists study changes as a result of new technologies and concepts, but brain science is interdependent with other socioeconomic and cultural contexts. One of the reasons that science changes is that society changes, and there is a relationship of mutual influence between science and other dimensions of culture.[8] As I suggested at the end of the introductory chapter, the way scientists understand and speak about the brain at any given point in history resonates with vocabularies and patterns of arrangement in other areas of social life, including culture, politics, and economics. I use the word resonance to gesture toward the relationship between science and other cultural assemblages for two reasons. First, it avoids reducing this relationship to a one-way causal relationship in which scientific knowledge produces social change, or cultural change transforms scientific practice. Resonance leaves room for thinking about this relationship in more complex ways that acknowledge the interdependency of science and culture. Just as the predominant lexicon and available conceptual apparatuses of the day inform scientific articulations, social patterns are conditioned by the linguistic and nonlinguistic practices of science. On a deeper level, the very distinction between science and culture is troubled by their mutual constitution. One of the themes that emerges in the following case studies is that what counts as science and what counts as culture or politics are constantly changing as a result of rhetorical activities. The second reason I prefer the word resonance is that it suggests a certain similarity of form, but not an identity. I do not believe that there is a single principle governing each historical moment, determining everything from scientific practice to economic policy. The word resonance is intended to indicate shared patterns but avoid reductionist tendencies.

Two brief examples illustrate this notion of resonance. First, in 1855 Herbert Spencer wrote that no serious physiologist could deny that specific parts of the brain served distinct functions, because "localization of function is the law of all organization whatever," and "separateness of duty is universally accompanied with separateness of structure."[9] For Spencer, phrenology's veridicality derived not solely from empirical observation but also from broader cultural assumptions about the way in which the world was organized. These assumptions—in this case, that literally everything, whether biological or social, is organized into discrete structure-function combinations—provided a lens or paradigm through which Spencer interpreted empirical data and understood the mind. A more recent example is the concept of networks, which is widely taken as a near-universal organizational principle in much the same way Spencer describes localization of function in the nineteenth century. In Restak's description at the beginning of this chapter, the New Brain is arranged as a network, consisting of dispersed elements that interact and combine across distances. This concept of networks is also commonly used today to describe characteristic forms of social, political, and economic arrangement.[10] There is a resonance or parallel between the dominant ways we understand and speak of the biological body and the ways we understand and speak of our political or social body.[11] Michael Hardt and Antonio Negri note the historical persistence of these parallels when they describe how models of sovereignty have often been expressed by way of analogies between social and biological bodies. In the Hobbesian model, for instance, each body has "a single subjectivity and rational mind that must rule over the passions of the body."[12] This statement was in its time truthful in the sense that Foucault speaks of—a legible statement recognized and validated as veridical. The same statement could be applied to physiology and the political function of rule. Just as biological bodies were understood as being ruled by a unified central control (the mind), the state required a central authority to command the members of the political body.

Today, as Restak's description of biological networks illustrates, the body is widely understood as something that lacks a transcendent mind substance or unifying center of subjectivity. The brain is conceptualized as "a chemical event or the coordination of billions of neurons in a coherent patterns," and the "swarm" makes decisions.[13] As Hardt and Negri suggest, this conceptualization also applies to predominant ways of theorizing contemporary political and social organization. Government takes place not so much through the dictates of a centralized authority, but through various combinations and coordinations of dispersed networks of rule. At this stage, I do not want to engage the question of whether this form of rule is more or less desirable than alternatives. Instead, my purpose is simply to outline the parameters of this notion of resonance that suggests a complex relationship between the science of the brain and the dynamics of society. In later chapters, I look more closely at the different forms of rule associated with these different neuro-social assemblages.

I adopt the terms "diachronic discontinuity" and "synchronic continuity" to refer to these two premises, though with some reluctance and recognition of the need for qualifications because they risk the implication that contemporary neuroscience occupies an epoch or totalizing episteme completely sundered from historical precedents. To guard against this tendency, I emphasize that I do not invoke diachronic discontinuity and synchronic continuity as determinant characteristics of the world, but rather as the longitude and latitude of the particular lens through which I seek to map the neuroscience configuration. The story of the rhetorical brain could be told in other ways to illuminate different features of its operation. My own perspective tends to highlight the singularity, or distinctiveness, of the New Brain, the ways in which it is historically contingent and caught up in contemporary sociopolitical contexts.

Mapping Structure and Function: From Phrenology to Plasticity

The history of brain science can be told as a history of localization theories and methods.[14] What scientists and historians call the localization hypothesis takes many different forms, but its core doctrine is that different human functions, including motor skills, cognition, and emotion, can be traced back to specific brain structures.[15] While localization is generally acknowledged as the predominant paradigm guiding neuroscience from the nineteenth century until very recent decades, questions and controversies surrounding theories of cognitive localization have shaped the history of brain science for centuries and continue even today.[16] The localization hypothesis holds that the brain is naturally divided into discrete structures, regions, or, in the language of neuropsychology, modules, and that each region is responsible for carrying out clearly specified functions. To speak of a hypothesis in the singular is somewhat misleading, as localization theories vary widely according to how they divide brain structures and functions and how they conceptualize structure-function relationships. Over the centuries, scientists have devised a variety of different methods designed to "map" the brain by marking out the brain's regions and correlating them with their particular functions. These maps vary widely due to multiple scientific and extra-scientific factors, including changes in both the methods that scientists use to visualize the space of the brain, and the ways in which functions themselves are specified.

Brain Mapping and Methods of Visualization

Since localization became a widely accepted assumption of brain science in the nineteenth century, brain mapping projects have tended to follow a specific trajectory. The dominant trend since the dawn of the "era of cortical localization" has been, Stanley Finger writes, to "divide the major brain parts into progressively smaller functional units."[17] This trajectory is directly tied to changes in technologies for accessing the brain and in scientific habits of vision.[18] As brain mapping tools are able to visualize the human brain at more minute levels of spatial and temporal resolution, theories of brain function tend to correspondingly focus on increasingly microscopic elements and processes. From one angle, even the distinction between localization and distribution can be seen as a threshold of scale. For instance, in his criticism of localization theories, William Uttal argues that equating mental processes with different regions of the brain has little explanatory value. In fact, he argues, the "proper level of equivalence between the components of the brain and mind are to be found at another, more microscopic level," which is so complex and interactive that it cannot be adequately accounted for in the languages of modularity or regional specialization.[19]

Even theories that take localization for granted as a basic premise produce very different maps of the brain, depending on their particular modes of visualizing brain "territory." For example, in the early nineteenth century, phrenologists believed that brain functions could be mapped onto structures discernible in the shape of the skull. Much more recently, Wilder Penfield and Lamar Roberts's brain mapping endeavor received a much greater degree of scientific acceptance than did phrenology. Their famous depictions of exposed cortices overlaid with numbered tickets illustrate the paradigm that brain functions can be mapped onto specific locations of the wet brain itself. In these two cases, there is variation in how the very concept of brain structure is conceptualized. For phrenologists, brain structure was accessed by way of visual and tactile knowledge of the skull. For Penfield and Roberts, brain structure was not correlated with the shape of the skull, and the brain itself had to be directly accessed and made visible for scientific scrutiny. To some degree, variations in brain mapping paradigms are conditioned by the available technologies. The ability to visualize the wet brain enabled the changes in how the brain was known and conceptualized. Likewise, contemporary brain mapping technologies condition recent changes in how the relationship between human function and brain structure are mapped and conceptualized. Digital technologies condition the digital languages and concepts that are used to speak of and understand the brain. In short, there is an association between the visualizing apparatus (the technology through which the brain is accessed) and rhetorical and epistemological dimensions of brain science.

As functional imaging technologies like PET and fMRI grew in prominence as noninvasive tools for accessing the brain in the 1980s and 1990s, they reinvigorated a centuries-old scientific quest to map the brain by localizing human functions in specific areas of the brain.[20] Because these new imaging technologies could visualize the brain at ever-finer spatial and temporal resolutions, they provided persuasive evidence for those interested in tracing various human functions—sexuality, reading, thinking, criminality, and so forth—to identifiable brain structures. In more recent years, however, there has been what Uttal describes as a "major sea change from discrete localization to a distributed network approach," even though localization continues to exert considerable influence.[21] Uttal writes, "We are in the midst of a conceptual revolution of the meaning of the beautiful PET and fMRI brain images."[22] Because of new insights about imaging data, neuroscientists have moved considerably away from "constrained regional localization" to embrace a "dynamic and ever-changing distributed neural system approach."[23] When the brain is visualized "in time" and seen as interactive processes, it is more difficult to demarcate clear spatial boundaries, and brain organization is understood as graduated and continuous rather than strictly bounded. Despite these trends, localization is far from dead, and many contemporary theories advocate a modified localizationism—for example, where cognitive functions are produced by the interactions of "functionally specialized brain loci," a change that only shifts from "one specialized locus to many, each of which has a specific role to play."[24] At the far end of the connectionist side of the spectrum, however, "a more holist theme is being played out," one that eschews functional specialization altogether and sees various elements as playing different roles for different functions, all "so tightly interconnected that the individual parts cannot be disentangled from the whole" without losing their nature.[25]

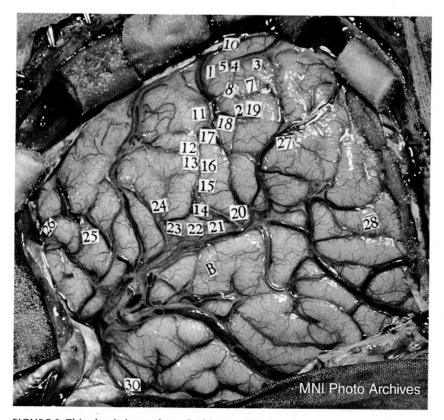

FIGURE 3 This classic image from the history of brain science suggests a one-to-one structure-function relationship. Penfield and Roberts exposed the brains of their epilepsy patients and then stimulated the brain to observe responses. Through this method, they "mapped" different functions onto specific cerebral locations. The original caption reads, "The dura has been turned back to expose the cortex. The left temporal lobe is seen below the fissure of Sylvius. Compare with Fig. VII-5 and with the Frontispiece. The numbered tickets, dropped on the surface of the cortex, indicate points of positive response to electrical stimulation."

Reproduced from Wilder Penfield and Lamar Roberts, *Speech and Brain Mechanisms*, Figure VII-2, p. 114, © 1959 Princeton University Press, 1987 renewed PUP. Reprinted by permission of Princeton University Press, and with permission of the Literary Executor of the Wilder Penfield Estate.

Brain Mapping and Vocabularies of Function

In addition to discontinuities in modes of visualization, the very ways in which brain functions are conceptualized are highly variable. In *The New Phrenology*, Uttal identifies this variability as the primary obstacle to the scientific quest to understand the mind. He provides copious data documenting how taxonomies of mental processes change across history and in different social and cultural contexts. He writes, "There has been and continues to be a plethora of classification systems of cognitive processes, and none has stood the test of time."[26] Moreover, there is no "progressive" pattern among classificatory schemes, no convergence or trend of moving toward a unified, coherent, and stable system.[27] For example, one early taxonomy includes "Discriminability" and "Character" as cognitive functions, yet these terminologies are not used in present-day classification systems. Uttal draws these two sources of discontinuity together when he describes how visualizing apparatuses have a close relationship to the types of classification terminologies a particular science finds compelling. With the emergence of digital imaging technologies, a new taxonomy is emerging, one that draws on the categories of mental function prominent in contemporary culture (Uttal gives as examples "unpleasant emotions" and "executive processes").[28]

In tracing divergent schemes of classification, the ultimate point is that scientists draw from terminological reservoirs of the contemporary culture to craft their classification schemes, and aside from any concerns over the accuracy of the technologies themselves, there is no objective way to determine what cognitive and emotional functions *are* in essence, to identify, label, and measure them in any reliable fashion. In other words, even if the measuring devices are guaranteed, the status and very essence of what is to be measured is dependent on cultural and social factors. The faculties of phrenology have given way, for instance, to *Diagnostic and Statistical Manual* (*DSM*) categories such as attention-deficit disorder, generalized anxiety disorder, and premenstrual dysphoric disorder. The numerous disagreements generated by ever-shifting categorizations of human function are ongoing, as evidenced by contemporary disputes over the forthcoming fifth edition of the *DSM*, the "bible" of psychiatric diagnosis. Proposals such as the inclusion of a new disorder, "Temper Dysregulation Disorder," and the removal of "dementia" from diagnostic phrasing, are a matter of public controversy and suggest the cultural stakes at play in attempts to categorize mental functions.[29]

Although the methods and concepts associated with localization theories are discontinuous, one shared trajectory is a commitment to a materialist premise that virtually all mental functions are products of the biological brain. Mapping the brain is thus consistently articulated as a way for science to gain complete knowledge about human nature. In *Mapping the Mind*, Rita Carter writes that when the brain is fully mapped, "the resulting description will contain all there is to know about human nature and experience."[30] To speak true regarding human nature is to speak in the language of the biological brain, articulating emotion and cognition in terms of neural geographies. For example, Jeffrey Schwartz and Sharon Begley describe the modern "triumph of materialism" in terms of the truth-value of biological language: "When you trace depression to activity in a circuit involving the frontal cortex and amygdala, you have—on the whole—explained it. When you link the formation of memories to electrochemical activities in the hippocampus, you have learned everything worth knowing about it."[31]

The assumption that human thought and personality can be fully explained in biological languages or accounted for with recourse to material nature is common to the localized brain of phrenology, the intermediate brain, and the plastic brain. Despite their shared proclivity for biological explanations, these three brains are articulated in very different ways—they have different capabilities, different properties, and, most important, different cultural effects. Biological accounts of the brain are more than scientific theories; they are culturally laden discourses that have substantial implications for how a culture conceptualizes identity. By identity, I am referring to the general cultural understanding of what a person is capable of and what relations are possible between the individual and society. Themes associated with this general conception of identity include how individuals acquire their social role, and whether a social role is fixed or plastic, and whether it is innate or acquired. To simplify the notion of resonance a great deal, if the brain is conceptualized as a plastic entity capable of substantial change due to various social stimuli, then individual identity and society as a whole are conceptualized as similarly plastic and malleable. Likewise, if brain biology is viewed as static, then both an individual's social role and social organization as a whole are thought to be more fixed and less amenable to dynamic restructuring. Put simply, the way that a culture understands the function of the brain has a certain relationship to the way a culture understands identity and social structures.

Phrenology's Brain

Phrenology is often invoked as a reference point for contemporary brain mapping endeavors, and attitudes toward phrenology provide a useful (although not definitive) marker for assessing a particular theory's place on the localization-connectionist spectrum. For example, in *Mapping the Mind*, Carter writes that Franz Gall, the founder of phrenology, was "prescient" in his ideas of "a brain made of functionally discrete modules."[32] Connectionists are more dismissive: Restak calls phrenology "pure hooey," and Goldberg describes it as "a false start" that belongs in the "prehistory" of neuroscience.[33] Moreover, because "Americanized" phrenology was "a cohesive cultural factor, uniting phenomena that might otherwise seem disparate," it provides a helpful point of comparison for the argument that the contemporary rhetorical brain is not a strictly scientific phenomenon, but rather deeply imbricated in political, social, and economic currents.[34]

Although widely considered a "pseudoscience" today, phrenology garnered considerable respect from scientists and intellectuals in the nineteenth century.[35] In addition, it wielded considerable force in American popular culture, shaping common understandings of the capacity to reform individuals by cultivating their innate potential, and to improve society by arranging individuals according to their traits and abilities. By the middle of the nineteenth century, the vocabularies of phrenology had been widely disseminated and were spoken regularly by individuals to articulate their struggles and shortcomings, as well as their dreams and aspirations.[36] Gall's phrenological science, later popularized by Orson and Lorenzo Fowler, is perhaps the most notorious attempt to localize human functions in discrete brain regions.[37] Gall viewed his phrenological method as a true science, grounded in empirical access to material nature. John von Wyhe describes Gall's ideology thus: "Nature, with a capital N, was the key to an authoritative voice in philosophical questions about human nature."[38] The order of "Nature" was for Gall necessarily visible, and the truths of human nature revealed themselves in the shape of the skull. Gall posited a correspondence between the strength of a mental faculty and the amount of cerebral space devoted to the faculty, and he believed that the amount of cerebral space would in turn be represented by protrusions of the skull. The Fowler brothers, who played an important role in making phrenology fashionable as a social reform method, infamously popularized Gall's science. I draw primarily from the Fowlers' example to illustrate how popular versions of phrenology articulated the relationship between the brain and identity.

Today, phrenology is most often associated with the diagram of the head on which the phrenological categories are mapped, "the most persistent vestige of the phrenological theory."[39] These diagrams depict a head mapped out into discrete regions, with each region clearly labeled as responsible for a determinate function (emotional, cognitive, or sensorimotor). Functions and structures correspond exactly in an analogical relationship that can be mapped onto the static space of the skull. Functions are precisely located in clearly demarcated regions, and the physical size of the brain region (as reflected in the shape of the skull) was thought to correspond directly to the strength or power of the correlated function. Nature operates, the Fowlers write, by a logic of "the most perfect reciprocity," such that when Nature puts forth "*power* of function, she does so by means of power in the *organ* which puts it forth."[40] The correspondence between organic structure and function is "fixed and absolute—is universal, not partial—is a relation of

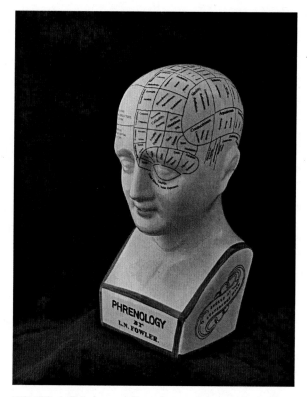

FIGURE 4 This is a replica of one of the Fowlers' phrenology heads, outlining the different faculties that were assigned to specific locations of the brain based on the size and shape of the skull.

Paul Thornton/Pt. Pocket Photography

cause and effect, and governs every organ and function throughout universal life and nature."[41]

In many ways, phrenology favored determinist understandings of human nature with its view that "the parameters of mental and physical ability are largely drawn by heredity, and one could only make the best of endowments received at birth."[42] Despite these tendencies, the Fowlers are credited with building a "practical phrenology" designed to improve individuals' abilities to navigate life and achieve happiness. This practical phrenology was motivated by a "doctrine of perfectionism," which believed that the mundane chores of life could be practiced in such a way as to "cultivate" and "restrain" different faculties. The Fowlers produced copious discourses on child rearing and family arrangement, suggesting that deliberate, targeted programs of cultivation and restraint could have especially powerful effects on children's development. Even for adults, biology determined character, but if individuals became knowledgeable about their own biological composition—and hence their natural characters—they would be better able to cultivate and restrain habits of existence appropriate to their own makeup.[43] According to the Fowlers, phrenology could determine, for instance, which profession an individual should pursue, and which should be avoided. Physicians, for example, require large "Perceptive Faculties" and large "Constructiveness," among other traits. They need "a large head, to give them general power of mind," and phrenology "will predict, in advance, whether or not a boy will succeed in this profession." Likewise, some individuals will make better merchants than others. Why is one better than the other? The Fowlers respond, "Phrenology answers this by pointing out the constitutional differences, and showing who is, and who is not, adapted to mercantile life."[44] As Madeleine Stern documents, the Fowlers were so successful in their vocational guidance and employment counseling that advertisements for help often required applicants to present results of a phrenological analysis before they would be considered.[45] In an 1852 editorial, Horace Greeley insisted that railroad accidents could be reduced if only conductors were required to undergo phrenological testing to determine their fitness for the position.[46]

Phrenology exerted considerable cultural force, especially in the way it framed the link between identity and work. The phrenological heads visually condensed a certain politics of labor characterized by themes of spatial fixity; as Stern writes, "There was a place for everything, and everything was in its preordained place."[47] Just as each human function had a fixed place or structure to which it corresponded, each individual had an ideal place in the social organization of work. Moreover, there was a direct correspondence between the functions and structures of one's brain and the functions and structures that one was capable of fulfilling and occupying in the context of work. The structures of one's head were thought to be an unerring guide to whether one was suited to a career as a physician, merchant, mechanic, or editor. A predominant or deficient temperament or characteristic could be linked in perfect reciprocity with a physical structure, because the temperament gives "a particular form to the body—shape being its index."[48] These analogical relations of structure and function allowed individuals to be precisely categorized into the various social paths that represent the most appropriate "fit" for their particular temperament. Individuals were urged to identify the noted character traits of both themselves and others and to recognize that there is a "marked coincidence between the phrenology and the character" of the person under observation.[49]

Once the truth of phrenology had been ascertained by recognizing this logic of coincidence, individuals could use this knowledge in practical fashion to ascertain appropriate channels for social and personal fulfillment, especially in the context of a career. The Fowlers' *Practical Guide* includes a tool to determine one's most suitable trade or profession based on the "fit" determined by their character, as evidenced through physiology. The relation between character, phrenology, and profession is one of correspondence or reciprocity that is largely determined, or preordained. Although biology, and hence character, could be modified to some degree through cultivation and restraint, practical phrenology worked to achieve social perfection by identifying character through biological examination, and then situating that biological composition within the social niche where it best fit. Once an individual is determined for a suitable trade or profession, that trade or profession becomes an essential quality of his or her character.

The scientific status of phrenology has been much contested, and references to present-day brain imaging as the "new phrenology" are received with varying degrees of acceptance and hostility. Whatever phrenology's scientific standing, it remains an important element in the history of attempts to localize mental functions in brain structures. Further, it provides an illustrative example of how scientific conceptions of the brain are implicated in broader socioeconomic arenas. In the case of phrenology, the articulation of the brain via particular visual and verbal apparatuses resonates with a contingent politics of labor and identity that suggests a one-to-one relationship, or "fit," between each worker, the worker's social placement, and the worker's economic function. In the next section, I turn to Damásio's *Descartes' Error* to examine how an intermediate view frames the brain as the ultimate source of everything human. I return to phrenology's politics of labor in the final section of this chapter in the analysis of Begley's plastic brain, which illuminates a resonance between brain plasticity and the "flexibilized" labor politics of contemporary capitalism.

Descartes' Error: Articulating Mind, Brain, and Body

The mantra in contemporary culture that the mind and self are nothing but effects of the brain echoes phrenology's veneration of the brain and material nature as the determining forces in human life. As a recent *National Geographic* cover story asserts, the mind is simply "what the brain does." From the cover, a contemplative-looking monk wearing an electrode cap looks calmly at the viewer. The layout preceding the feature story depicts the same monk, legs and arms crossed, with a meditative expression, with the words "the mind is what the brain does" in deep hues against a black background. The monk evokes connotations of religious tranquility, metaphysical mystery, and a mystical transcendence. The image of the monk under observation suggests that brain imaging can access not only neurochemical patterns but also the very substance of spirituality. The brain is a sacred organ; no

FIGURE 5 This figure was featured on the cover of *National Geographic* in March 2005. The monk is wearing an electrode cap that might be used for an EEG brain scan. In the cover story, the images of the monk are featured next to captions such as "the mind is what the brain does," suggesting that all the spiritual activities associated with the monk can be traced back to the biological brain and measured through imaging technologies.

Cary Wolinsky/*National Geographic* Stock

other part of the body promises to disclose such deep truths about the fundamentals of human existence.

This example is representative of a trend in popular and scientific discourse whereby virtually all aspects of human existence, including those traditionally located in the sphere of "mind"—cognition, emotion, and personality—are being defined in the terminologies of material biology. In short, brain rhetorics articulate the core of individual essence as an effect or product of the biological brain. If the brain is the seat of human essence, then science is capable of revealing the entirety of human nature. Gerald Edelman claims, for example, that developments in neuroscience are "a prelude to the largest possible scientific revolution," which will result in knowledge of "how the mind works, what governs our nature, and how we know the world."[50] Similarly, Francis Crick predicts that if we "press that attack," science will come to explain "*all* aspects of the behavior of our brains," and "we shall be led to a clear understanding of our true nature."[51]

This combination of materialist thematization (everything can be reduced to events of biological brain) and optimism regarding the perfection of scientific knowledge finds bold formulation in the influential book *Descartes' Error*. António Damásio is recognized as one of the world's leading neuroscientists and a pioneer in scientific research into the neurobiology of emotion. An endowed professor of neuroscience at the University of Southern California and director of the Brain and Creativity Institute, Damásio has written a number of books that, like *Descartes' Error*, are aimed at a mixed audience of neuroscientists and laypersons. In 2005, a tenth-anniversary edition of *Descartes' Error* was published. The book has been translated into thirty-three languages and is often used in college and graduate seminars in neuroscience and neuropsychology. In the book, Damásio argues that the "self" is nothing but physiological processes, "a repeatedly reconstructed biological state."[52] Damásio insists that the "self" is not the product of the brain alone, but rather of the brain-body ensemble acting in coordinated fashion.

Damásio's emphasis on biological foundations is accompanied by claims that neuroscience has overcome age-old binary oppositions and laid the foundation for a genuine materialist monism. As his title suggests, Damásio's book is dedicated to denouncing Descartes' error, identified as both "the abyssal separation between body and mind" and the assumption that thinking and awareness, rather than biology, constitute the real substrates of being (*DE*, 248–250). For neuroscience, claims to a monistic ontology rest on fusing the two realms sundered by Descartes and authorizing biology as the only really existing substance. Mind exists only insofar as it is conceived as an effect or simulation caused by primary biological activity. This conception of mind and consciousness as effects of physiology is predominant in both scientific and philosophical theories. The philosopher Daniel Dennett, for instance, has argued for a materialist view of human thought along similar lines, arguing that the mind and our sense of "I," or unified subjectivity, are simply epiphenomena, basically illusions produced by dispersed physiological processes.[53] Similarly, for Damásio the idea of a core essence—mind, soul, or subjectivity—that unifies an individual and guarantees one's identity is an illusion or an effect produced by the interaction of different brain regions acting in concert. The perceptions that we are unified creatures or "selves" are tricks of timing, the result of activity in different regions occurring simultaneously (*DE*, 95).

Damásio develops his theory of human nature through the tales of individuals who have suffered brain injuries. Although Damásio also relies on contemporary technologies, including imaging, to make his case, his discussion of injured patients has much in common with the "lesion method," a way of studying the brain that has been used for centuries, especially before noninvasive imaging technologies were widely available. In the lesion method, scientists observe an individual with an injured brain and use the injury to account for any deficits or abnormalities observed in the patient's behavior. The part of the brain that suffered damage is then assumed to be responsible for the normal human function associated with whatever type of abnormality or deficit was observed. Thus, if a patient with a damaged frontal lobe exhibits impaired reasoning, a brain scientist using the lesion method would assume that the frontal lobes are responsible and necessary for proper (normal) reasoning activities.

Both the lesion method and localization efforts tend to manifest tendencies toward determinist accounts of mind-brain interaction. In other words, when a certain identifiable brain region or regions are associated with a particular cognitive function, it is far more common to understand causality as running from the brain to the mind, and not the other way around. This is seen, for example, in "the mind is what the brain does" statements, which suggest that the brain is the cause of all mental events. Mind, in these discourses, is shorthand for all the intangibles of human existence, including personality, thought, and social interaction. Damásio's book illustrates both the ways in which localization rhetorics manifest these determinist tendencies, and the tensions that inhabit determinist explanations of the brain. In the next section, I describe how the relationship between localization discourses and determinist claims operates in popular neuroscience; after that, I address how attempts to define the brain as the source of the mind and self ultimately contribute to the articulation of a sacred brain, a construction that is responsible, at least in part, for the cultural force of the rhetorical brain.

Damásio's Intermediate Localizationism

The story of Phineas Gage is a critical centerpiece of many popular neuroscience texts, including *Descartes' Error*. In the popular version of the story, Gage was a construction foreman known for his leadership, integrity, and commitment to his work, the "most efficient and capable" among the men laying railroad track across the state of Vermont (quoted in *DE*, 4). In 1848, Gage was in an accident that drove an iron rod through his head and damaged part of his brain. Gage survived the accident with his cognitive abilities apparently intact, but his life was never the same: the conclusion of friends, family, and medical professionals was that "Gage was no longer Gage." After recovering physically from the trauma, Gage's personality was almost entirely transformed from a capable leader to a man "fitful, irreverent, indulging at times in the grossest profanity" and unable to restrain his "animal passions" (quoted in *DE*, 8). Gage's riches-to-rags example is commonly invoked as an early lesion study proving that the brain comprises discrete, specialized units. Gage was still able to walk and talk after his accident, but his personality deficits are taken as proof that the parts of the brain damaged are responsible for mature social reasoning.

For Damásio, Gage's story provides proof for two related claims: First, all aspects of human function, including abstract ones like social reasoning and mature forethought, can be traced back to localized processes in various brain regions. Second, every normal, complex function depends on multiple regions acting at the same time and cannot be reduced to a single center. Specifically, Damásio argues that attempts to locate cognition in a single region are faulty because they fail to account for the role of emotion in reason. Regions responsible for emotional processing contribute to processes of reason just as much as those areas of the brain responsible for cognitive operations. In essence, Damásio advocates what Uttal has described as "an intermediate position" that sits somewhere between extreme localizationism and distributed connectionism.[54] Another way of thinking about Damásio's view is as a sort of pluralized phrenology, where instead of correlating functions with a single brain region, each function is linked to multiple regions acting simultaneously. I situate Damásio as an intermediate localizationist for three specific reasons: First, he views spatial location as the determining factor for an element's function. Second, Damásio argues that specialization is firm, and that an element cannot interchange its function for a different role. Finally, although Damásio does argue that the brain can change—and, indeed, some portions of the brain are more dynamic—change is constrained by location and highly restricted by neuroanatomical structures. In many ways, Damásio's theory of the brain is fairly continuous with the phrenological view. Damásio himself claims that the phrenologists have been validated, as "brain specialization is now a well-confirmed fact" (*DE*, 15). Moreover, like the phrenologists, Damásio holds that functions are localizable in a fairly straightforward way, although the technologies of brain mapping have evolved (scientists no longer look to the skull to access brain structure) and the goal is now to identify multiple modules for each function, rather than a single region.

Despite affinities with phrenology, Damásio persistently preempts any accusations of phrenological thinking. Damásio claims to avoid the "trap" of phrenology by rejecting a simplistic notion of brain centers. As he writes, "There are 'systems' made up of several interconnected brain units . . . and these systems are indeed dedicated to relatively separable operations that constitute the basis of mental functions" (*DE*, 15). Depending on where they are located in a particular "system," units "contribute different components to the system's operation and thus are not interchangeable" (*DE*, 64). Location determines a region's contribution in producing a particular function, and each contributing region is necessary for the function—one region cannot, for instance, substitute for another in the case of a serious injury like Gage's. About this view, Damásio insists, "there is nothing phrenological" (*DE*, 99). Damásio's repetitive insistence on distancing his view from phrenology is perhaps derived from his recognition of the appearance of similarity between the two views—recognizing this potential perception of commonality, Damásio takes extra care to deny this association and separate his own view from one that is today widely disparaged as pseudoscience.

Although Damásio distinguishes his theory from the strict localization theories of the phrenologists, he also dismisses connectionist understandings of mental function. Connectionists, according to Damásio, see that the brain comprises multiple complex circuits and connections; but because the task of mapping out these circuits and connections seems so daunting, connectionists essentially give up and just argue that the brain is "one big morass and all functions emerge from this mass" (*DE*, 29). In fact, Damásio insists, connectivity is limited and highly constrained by location: "Pick a few neurons in the cortex or in nuclei . . . and you will find that each neuron talks to a few others but never to most or all of the others." In fact, he continues, "many neurons talk only to neurons that are not very far away" (*DE*, 29). Thus the brain is interconnected to an extent, but spatial location is the critical factor determining the contribution any one element will make to the overall system or function. Differentiation and specialization are products of spatial location, which is for the most part, barring injury, determined or fixed by natural biological processes. When a specialized area of the brain is damaged, the consequences are often permanent: "The brain is not one big lump of neurons doing the same thing wherever they are. The structures destroyed in Gage and Elliot [a patient who suffered brain damage caused by a tumor] happened to be those necessary for reasoning to culminate in decision making" (*DE*, 39). Damásio explains the interaction of spatially distributed regions not through connectionist theories, but rather as a "trick of timing": "Probably the relative simultaneity of activity at different sites binds the separate parts of the mind together," thereby creating an illusion that synchronous activity occurs in the same place or "theater" (*DE*, 94–95). Thus different regions are discrete modules that work together not so much because they are connected to one another, but because they become active at the same time.

Determinism, Pathology, Localization

In popular neuroscience, there is a tendency for certain rhetorical elements to cohere. Stories of brain injury or brain pathology are often used to illustrate localization theories, and these stories are often interpreted to contain a determinist moral. By determinism, I am referring to a rhetorical practice or pattern of articulation, in which the brain is assigned causal responsibility for human functions, or as the case more often tends to be, human malfunctions. By examining the coherence between tales of brain pathology, localizationist explanations, and determinist rhetorics, I do not aim to trivialize or negate the basic idea that brain pathology can cause cognitive and personality deficits. In other words, I am not arguing that these explanations are false—again, my goal is to bracket the question of truth or falsity and attend to rhetorical patterns in popular neuroscience in order to assess their cultural effects. Moreover, most of the popular texts I examine are not uniform in their commitment to localization and determinism. Damásio's book, for instance, includes many examples of what I am calling determinist rhetoric, but *Descartes' Error* also contains explicit attempts to validate notions of free will, choice, and human responsibility. Determinism and optimization are not mutually exclusive positions in popular neuroscience; rather, determinism is the flip side of optimization, and both play important roles in brain culture. Most popular neuroscience rhetorics incorporate elements of each, and my task is not to decide between the two but to trace the associations between localization and determinism, and connectionism and optimization, in order to shed light on why popular neuroscience is such a pervasive and compelling language for articulating multiple areas of human life.

Damásio's tales of pathology are entwined with determinist elements, which are most visible when he draws moral conclusions from the science of localization. Damásio warns that ignorance of the fact that distinctive parts of the brain are responsible for mental functions leads to unjust social policies, including incarceration and the death penalty, which punish individuals for actions ultimately out of their control (*DE*, 19). Even individuals with no apparent brain pathology can suffer lapses in ethical reasoning due to blips in brain processes: we should, Damásio writes, work to reconsider even "our own responsibility when we 'normal' individuals slip into the irrationality that marked Phineas Gage's great fall" (*DE*, 19). It is deeply problematic to "blame" individuals for character flaws, "defective emotional modulation," and "lack of willpower" when the underlying cause of these things is brain disease (*DE*, 40). For Damásio, brain disease is a broad explanation that can apply to an enormous range of undesirable behaviors and attitudes, even those that are not accompanied by clearly visible or identifiable pathology.

Human "disease" includes pretty much all varieties of human suffering: "the proclamation of bruised feelings, the desperate plea for the correction of individual pain and suffering, the inchoate cry for the loss of a never-achieved inner balance and happiness to which most humans aspire" are all medical issues, aspects of "human disease" (*DE*, 257–258). Although humans do have, in some cases, a capacity to resist the dictates of biology, it is also the case that there are "potions" within our brains and bodies "capable of forcing on us behaviors that we may or may not be unable to suppress by strong resolution" (*DE*, 121). Brain disease is, in Damásio's rendering, a "disease of the will" in that it renders human will ineffective.[55] The conclusion Damásio draws is that individuals should not be blamed or punished for negative behaviors and attitudes that result from these brain diseases.

There is nothing about the idea of location that necessarily leads to determinist conclusions. This rhetorical tendency is fed by a long history of localizationist theories that speak in metaphors of hardwiring and machines, which imply static structures. These theories tend to associate location with a sense of fixity, stasis, and natural determination that, in Doidge's words, "leaves little room for plasticity."[56] Further, there is nothing about determinism that necessarily requires a localization hypothesis. For instance, it seems just as possible to define the brain in terms of interconnected, distributed neural processes and assign these processes a causal role in producing undesirable behaviors (and this framing does circulate in popular neuroscience). For some reason, however, there is something about the ability to point to a definitive location or locations—saying in effect "*here* it is; here is the source"—that makes determinist explanations so compelling and lends them the air of truth.

The coherence of localization and determinism is not isolated to Damásio's book; in fact, it can be found throughout popular neuroscience. A 2007 cover story of *Time*, for instance, takes the most enduring questions of human being—the sources of good and evil—back to the brain. Good and evil, visually epitomized by Gandhi and Hitler, respectively, are featured against a brain scan image underneath the claim that science, through its investigations into the brain, is discovering the very sources of moral behavior. The cover suggests that human moral qualities are effects of brain function, and that there are specific areas or structures of the brain responsible for both moral and immoral capacities. An image inside the magazine depicts a diagram of the human brain, with specific regions highlighted to indicate where "moral decision making" occurs. The cover story substantiates this framing, referencing the infamous story of Gage and his transformation from "kindly to belligerent." According to *Time*, Gage's story kicked off scientific attempts to locate "the roots of serial murder in the brain's physical state," efforts that continue today. Brain scientists have found evidence linking specific neural structures—the amygdala, "a deep structure that helps us make the connection between bad acts and punishments," and the frontal cortex—to the capacity for moral behavior. When these structures are damaged, the result is aberrant behavior, as illustrated by Gage's cautionary tale.[57]

Carter, in *Mapping the Mind*, also engages the localization-determinism nexus. She cites a brain scanning study of forty-one convicted murderers that found that the majority showed "reduced frontal lobe activity," and that this "may severely compromise a person's ability to control their impulses."[58] She includes brightly colored brain scan images comparing a "normal" brain with the brain of a murderer, and she explains that violent criminals often exhibit dysfunction in critical areas of the brain. Carter comes to much the same conclusions as Damásio, arguing that it is "pointless and unfair" to punish perpetrators: "if brain mapping fulfills its promise of revealing exactly what is going on in rage-driven brains, it should also be able to suggest more appropriate means of dealing with them," namely, medical rather than punitive treatments.[59] These conclusions are not simply speculations of individual authors: they have real cultural impact. For example, a growing area in legal studies is "neurolaw," which examines, among other things, the way that brain images can inform courtroom testimony and judicial decision making across a number of different contexts. Some have argued that neuroimaging will eventually be used to "demonstrate the propensity for violence, the capacity to stand trial, as evidence of malingering, or to help establish or diminish the criminal responsibility of a defendant."[60] In "The Brain on the Stand," Jeffrey Rosen writes of imaging research that is working to "develop a deeper understanding of the criminal brain, or of the typical brain predisposed to criminal activity," leading to appropriate preventive and punitive measures.[61]

Violence and crime are not the only negative behaviors at stake. For example, in Louann Brizendine's CNN article "Love, Sex, and the Male Brain," she writes that men are more sexually aggressive than women because "men have a sexual pursuit area that is 2.5 times larger than the one in the female brain." This makes it "impossible for them to stop thinking about female body parts and sex." In this example, Brizendine relies on localization theories, assuming that there is a certain area or module of the brain responsible for sexual pursuit. Because brain images show that this structure is larger in males, it is assumed that men are determined or caused to engage in activities associated with sexual pursuit. This determining brain structure is "not likely to change" and should simply be accepted as a reality: "The best advice I have for women is to make peace with the male brain. Let men be men."[62] In this example, specific locations or regions of the brain are tied to distinctive functions, and this structure-function link supports determinist claims—in this case, men's sexual behaviors are attributed to forces outside their control in a classic (and disturbing) example of the "my brain made me do it" discourse. As these examples show, the languages of localization have considerable presence in popular neuroscience, even as the circulation of connectionist vocabularies continues to accelerate.

The deployment of images and localization discourses suggesting that negative human function—from an individual's temper tantrum to the collective evils of genocide—can be explained by the brain show that determinist explanations, or those that give causal power to biology, are attractive in our brain culture. Determinism is alluring precisely because it allows us to distance ourselves, as individuals and as collectives, from undesirable and harmful attitudes and behaviors. Moreover, it provides a way of making the inexplicable—human evil—legible and, at least in theory, manageable. If negative actions can be explained as an effect of brain dysfunction or disease, then perhaps error and evil in all their forms can be eliminated once and for all through scientific knowledge and medical treatment. In the concluding chapter of Brain Culture, I take a closer look at the utopian dreams that inhabit popular neuroscience in both its determinist and agency-centered instantiations. For now, I simply want to note that localization theories remain an important part of brain culture, in part because they provide explanations for human error and evil that offer at once the satisfaction of knowledge, a sense of distance, and the hope of cure and redemption for fallen humans.

The Stubborn Persistence of Subjectivity

Despite the cultural power of determinist explanations, the attempt to define all human subjectivity as a product of brain biology is a challenging enterprise that requires significant changes in the way biology itself is defined and understood. Despite his attempts to articulate a consistent materialism that moves beyond mind-body dualism, Damásio ultimately cannot avoid dualist or oppositional discourse. Damásio locates virtually all aspects of human existence within biological or material causation, but when he does this, he is forced to describe biology itself (and the brain in particular) as something that has will, agency, and subjectivity. In other words, Damásio does not reduce human subjectivity to the brain so much as relocate the qualities of mind, will, and agency onto the brain. The point is not to indict Damásio for failing to affirm a genuinely materialist monism, but rather that attempts to do so ultimately end up ascribing special properties to the brain, contributing to the construction of the brain as an object of cultural fascination and reverence. In popular neuroscience at least, accounts of human nature continually fall back on the categories of will, agency, and choice, whether those are defined as belonging to a subject or a group of neurons. Even if these assignations are metaphorical, their ubiquity in popular rhetorics plays an important role in shaping cultural attitudes toward the brain.[63]

As Damásio seeks to confirm a materialist monism by attributing processes formerly described in terms of an immaterial mind to the biological brain, his discourse produces new tensions between the brain and the rest of the body. At the core of Damásio's argument is the premise that neuroscientific research necessitates a shift in focus, from the cognitive or psychological dimensions of thought to the biological origins of thought, an attention to "why the brain's neurons behave in such a thoughtful manner" (DE, 251). Properties such as "thought" are attributed to neurons and conceptualized as fundamentally physiological, rather than mental or psychological, processes. The language of biology becomes the stopping point for explanations of human cognition, affect, and behavior—for Damásio, once a particular process or event can be described in the terms of neuroscience, it is epistemologically secured. Neuroscience is a way of "speaking true" in Foucault's sense, because it is privileged as the ultimate transparent language that can access the truths of human nature.

When the processes of mind are articulated in the language of material biology, biology itself becomes a dual essence of sorts. In other words, oppositions between mind and brain are transplanted onto the brain itself. Damásio's articulation manifests a covert dualism, illustrating how redefining the mind as an effect of the brain does not obviate the need for oppositions. Damásio writes, "Whatever questions one may have about who we are and why we are as we are, it is certain that we are complex living organisms with a body proper ('body' for short) and a nervous system ('brain' for short). Whenever I refer to the body I mean the organism minus the neural tissue (the central and peripheral components of the nervous system), although in a more conventional sense the brain is also part of the body" (DE, 86). In this illustrative passage, the mind-body dualism gives way to a brain-body opposition, as the brain is partitioned off from the rest of the body and given a special status. Only in a "conventional" sense is the brain part of the body; in terms of neuroscientific descriptions, the brain has a special status that cannot be reduced to the mute physiology associated with the passive body. Damásio is forced to structure his explanation of human nature in a form partially reminiscent of the very Cartesian dualism he is intent to overcome.[64] As Elizabeth Wilson points out, the brain becomes "an organ blessed with wondrous capacity," and the rest of the body is relegated to a subordinate metaphysical status.[65]

Whether these residual oppositions are theorized as an inherent product of grammar, the inevitability of metaphysics, or some combination of both is less important than what they show about how popular neuroscience frames the brain. In Damásio's account, the brain is still figured as a biological organ, but it is simultaneously a metaphysical vessel with attributes of will and consciousness. These metaphysical properties, or the intangible characteristics conventionally associated with mind, are folded into the meta-discourse of biology and spoken forth in terms of neurons, synapses, and regionally specific chemical events. By articulating the brain in terms of interactions between neurons, the brain is transformed from an inert object to a space of interactions that are simultaneously material and "intangible," making the brain an apt host for those elusive properties of human nature that formerly wore the signifiers of mind and will.

Damásio's use of communication metaphors to describe brain function aids his transformation of the brain into an active entity that contains the qualities of the mind, or human subjectivity. When Damásio describes the brain and body as distributed circuits or "assemblies" in which neurons "talk" to other neurons, even in their "relatively local circuits," he participates in a common mode of articulating brain functions via communication metaphors (DE, 29–30). These metaphors are not exclusive to localization accounts, and in fact they have wide circulation in discussions of connected, plastic brains. The consensus in contemporary neuroscience is that the brain is a space of communication, whether that communication occurs between neurons, regions, or chemicals. For example, the very first paragraph of a recent textbook contains the following: "The neuron is the communicating cell, and a wide variety of neuronal subtypes are connected to one another via complex circuitries usually involving multiple synaptic connections."[66] The neurobiologist Debra Niehoff ascribes even greater powers of communication to the brain than does Damásio, who holds that neural communication is relatively constrained by spatial location. Niehoff describes the communication processes of cells, particularly neural cells that are uniquely capable of communicating across long distances, as the equivalent of "telecommunications networks" that "permit even distant cells to share information, so that a biological society, regardless of its size or complexity, can coordinate the activities and regulate the social behavior of many of its members."[67] Cells are depicted as social creatures that "talk," and which must be "regulated" and "coordinated" by distant neural cells responsible for ensuring orderly function. These neural cells carry on conversations that go beyond the "everyday, the practical, or the essential," Niehoff writes. "The cells that make up our nervous systems describe sunsets, craft poetry, solve equations, remember birthdays, dream."[68]

In descriptions such as these, there is an equivocation between "communication" in the sense of Aristotle or Kenneth Burke—a uniquely human activity that requires intentionality or, at the least, some degree of motivated consciousness—and "communication" as a purely material process involving the transmission of chemicals and electrical signals. Neurons and other biological elements are implicitly endowed with qualities such as sociality, emotion, and will. Although these implications follow from descriptions that are designed to be purely metaphorical, their inevitability in materialist-inspired descriptions of biology shows how neuroscientific discourses refigure the brain as a dual entity that is the site for both biological processes and processes of subjectivity like will and mind. As Damásio's account shows, materialist narratives often smuggle in the categories of will and agency, disguised by a biological language that has been stretched to accommodate new meanings. Again, my point is not to draw attention to philosophical contradictions inhabiting popular neuroscience, but to highlight the ways in which popular neuroscience derives its rhetorical power from accounts that construct the brain as simultaneously cause and effect, both an agent that controls human nature and an object that can productively absorb the effects of deliberate human actions.

Of course, conceptions of the brain as a sacred, metaphysical object are not new, as historical attempts to locate the soul and other spiritual or religious entities in the brain illustrate.[69] What is interesting is that popular neuroscience sustains, albeit in modified fashion, the sense that the brain is something with a special metaphysical status. Popular neuroscience often sets science's abilities to access the uttermost reaches of the brain (and human nature) against mistaken views of history that saw the brain as divine or mysterious. For example, Crick denounces "earlier, prescientific ideas" about the soul, celebrating the "spectacular advance of modern science," which definitively proves that the so-called soul is simply biological events in the brain.[70] Edelman similarly derides primitive beliefs, including the "magic, vitalism, and animism [that] pervaded prescientific communication."[71] Like Crick and Damásio, Edelman's goal is to put the mind (and soul) back into nature and account for them fully as natural phenomena, wholly accessible to science and explainable in the languages of biology. *Descartes' Error* suggests, however, that despite this excitement over scientific progress and criticism of "prescientific" views, the brain is perhaps destined to remain, at least in the ways it is articulated and framed for public audiences, enigmatic. Despite his desire to account for the soul via the brain, Edelman defines the brain as "a *special* kind of process," depending on "*special* arrangements of matter."[72] Biological languages do not make the brain (or human nature) transparent to scientific knowledge; rather, they reshuffle categories and fold the elusive qualities of choice, will, and agency into the language of biology, attributing them to neurons or electrochemical processes (Crick, for instance, states that free will is "located in or near the anterior cingulate").[73] As I explain in the next section, theories of plasticity also contribute to the sacred aura of the brain in the ways that they frame the brain as both cause and effect of human thought and behavior; I then turn to Begley's book to unpack the persuasive appeal of plasticity and contextualize its resonance with broader cultural dynamics.

The Rhetoric of Plasticity

After the heyday of phrenology, other localization theories grew in prominence and many still enjoy a substantial degree of scientific consensus. Throughout the nineteenth century, support for localization grew thanks to critical breakthroughs by Paul Broca. Through the careful study and eventual autopsy of a patient whose speech was impaired, Broca determined that speech areas were located in the left frontal lobe. He concluded in 1861, "There are in the human mind a group of faculties and in the brain groups of convolutions, and the facts assembled by science so far allow me to state, as I said before, that the great regions of the mind correspond to the great regions of the brain."[74] Broca, like the phrenologists, expressed the relation between mind and brain as both causal (the brain causes mental operations) and based on a paradigm of correspondence or representation, where discrete locations are correlated with specifically demarcated functions. Throughout the nineteenth century, localizationists emerged as the scientific consensus against the rival holists, who tended to view the brain as something that could only function as a whole. A contemporary neuroscience textbook illustrates the degree of this widespread acceptance: "The nervous system is not a big blob; it is built from discrete units. If we can figure out how these units work, and describe the laws and principles of their interaction, then the problem of how the brain enables the mind can be addressed, and eventually solved."[75]

In recent years, these two assumptions—that the brain causes the mind, and that the brain is divisible into distinctive, specialized regions—have been supplemented, although not supplanted, by a different rhetoric that permeates popular neuroscience. The rhetoric of plasticity insists that there is what Daniel Goleman identifies as a "two-way street of causality" between mind and brain, and that determinist theories have been wholly debunked and are obsolete.[76] Moreover, the rhetoric of plasticity takes a different approach to explaining the structure and function of the brain. The brain does tend to comprise what can roughly be thought of as specialized regions or, more accurately, circuits, but these regions emerge from patterns of experience and are not prefigured by nature. Further, because individual experience is highly variable and always changing, these circuits are individualized and extraordinarily dynamic. To distinguish these two understandings of the brain's spatial order, Goldberg introduces the notions of a priori and a posteriori modularity. "Resultant" modularity is very different from a priori modularity: "It will exhibit less uniformity and a greater degree of individual differences and, most importantly, it will change over time as a result of the individual cognitive history."[77] These two views of modularity shape interpretations of brain scan images depicting regions of intensive activity. For the a priori view, the areas that are "lit up" on the scan image suggest a discrete region or regions responsible for the correlated function. For proponents of resultant modularity, these areas are "hubs" of high connectivity, and these areas are expected to change over time for the same brain and to vary across different brains.[78]

Training the Plastic Brain

Sharon Begley, a renowned science journalist whose columns have appeared in *Newsweek* and the *Wall Street Journal*, offers *Train Your Mind, Change Your Brain* as a primer on recent neuroscience research on brain plasticity as well as a practical guide to self-improvement.[79] Begley takes an extreme view of brain plasticity, holding that even motor and sensory functions—and not simply high-order cognitive processes—are produced by highly malleable circuits. Just as Damásio's book illustrates the coherence between localization discourses and determinist tendencies, Begley's book illuminates the relationship between connectionist languages and prioritizations of will, agency, and choice. After describing Begley's articulation of the plastic brain, I contrast Begley's plastic brain and its politics of labor with phrenology's politics of labor to show how different ways of articulating the brain are deeply imbricated in social, political, and economic spheres of cultural life.

The recent scientific attention to brain plasticity represents a shift from past dogma that the brain's organization was relatively fixed by an early age. Begley argues that this shift is largely driven by functional imaging technologies and their visualizations of active, dynamic brain processes. Imaging technologies such as PET, Begley writes, see "the whole brain, showing activity everywhere, and when the numerical readings are transformed into colors for easy reading, the spots of high and low activity practically scream at you" (*TYMCYB*, 91). The dynamic, brightly colored kaleidoscope of regions and circuits visualized by technologies such as PET stand in stark contrast to previous methods of seeing—and conceptualizing—the brain. The "prevailing dogma in neuroscience for over a century," that the brain's structures and functions are "static and stationary," is soldered to the qualities associated with the visual image of "a deathly white cadaver brain floating in a vat of formaldehyde" (*TYMCYB*, 6). In contrast, Begley ties theories of plasticity to the brightly colored, attractive productions of functional imaging technologies. To see the brain as static is to fail to see it as a living, functioning organ.

As Begley describes it, the New Brain is a series of elements that are not determined for any particular function—in other words, the relation between structure and function is not necessary but rather contingent. Specialization exists, but it is the product of experience rather than biological determination. All neural matter is, at an ontological level, the same—through experience it is forged into different types of matter with different capabilities and limitations. Begley explains, "No matter where in the brain a neuron lives, from the visual cortex to the somatosensory cortex, it is basically identical to neurons in any other neighborhood" (*TYMCYB*, 75). Neither biological essence nor spatial location determines a neuron's function, a view that is in sharp opposition to Damásio's insistence that brain regions are highly specialized and are not interchangeable. For the plastic brain, a neuron's function is shaped by experience, including environmental inputs and internal forces such as will, but it retains some degree of malleability, or the ability to take on new functions over the course of an organism's life span. Structure and function can be unlinked for motor and sensory as well as cognitive operations. For example, Begley cites studies showing how a brain's auditory regions can be recruited to visual processing tasks. Begley concludes, "Through neuroplasticity, the brain's structures are in no way struck with the career their DNA intended" (*TYMCYB*, 84). Brain maps correlating structures and functions might have some utility but should always be "printed in erasable ink," because the relationships they depict can be "completely overturned by the life someone leads" (*TYMCYB*, 10).

Begley engages two series of metaphors to describe neuroplasticity: those of real estate and labor, the latter exemplified in the previous statement about neural "careers." Neuroplasticity, she continues, "produces wholesale changes in the job functions of particular areas of the brain. Cortical real estate that used to serve one purpose is reassigned and begins to do another" (*TYMCYB*, 129). While the real estate images are spatial metaphors that resonate with discourses of location, they are primarily deployed to illustrate that locations are contingent and that the brain is, in Uttal's words, "spatially uncertain."[80] The real estate metaphors suggest an understanding of space as highly malleable in terms of boundaries and functions. Space can be "rezoned," and the boundaries can be redrawn as different topographical configurations emerge. Further, what a particular space is used for is not intrinsic to the space—it is a product of whatever elements (in the case of the brain, neurons) are currently occupying that space. As space changes hands and is occupied by new tenants, its function changes as well.

The career metaphors work with the spatial metaphors to amplify this sense of contingency. When an area's function is disrupted, it is "unemployed" and must be rezoned to take on a different function (*TYMCYB*, 97). An area of the brain can abandon "the career laid out for it from birth" and adopt new skills and tendencies in response to external and internal stimuli (*TYMCYB*, 99). Together, these two metaphors suggest Sir Charles Scott Sherrington's famous description of the brain from 1917, which Begley quotes: the brain is "an enchanted loom, where millions of flashing shuttles weave a dissolving pattern, always a meaningful pattern, though never an abiding one" (*TYMCYB*, 29).

Begley ties the understanding of the brain as an assembly of dynamic circuits to discourses of optimization. Drawing on Buddhist theories of the self, Begley correlates the plasticity of the brain with the malleability of human potential. Neuroscience concurs with Buddhism, defining a person as "a constantly changing dynamic stream" (quoted in *TYMCYB*, 13). Individuals are not set in their capacities; instead, they can optimize themselves by willfully manipulating the neural processes that compose the substrate of the self. Individuals can improve themselves by "enlisting the mind to change the brain," engaging in patterns of behavior that will "rezone" the brain's structure-to-function relationships. Begley continues, "A brain with no special ability in sports or music or dance might be induced to undergo a radical rezoning, devoting more of its cortical real estate to the circuitry that supports these skills." Regardless of an individual's genetic or "natural" biology, the opportunities for self-improvement are limitless: with appropriate mental techniques, Begley speculates, a suspicious and xenophobic brain can be improved into a compassionate and altruistic brain. Individuals can train themselves "to be kinder, more compassionate, less defensive, less self-centered, less aggressive, less warlike" (*TYMCYB*, 24–25). This training depends not on drugs or external causes, but rather on internal, mental attitudes and forces such as "will." Individuals have the ability to *decide* and "willfully direct" "which moral capacities emerge and which do not, which emotions flourish, and which are stilled" (*TYMCYB*, 241). Begley makes explicit that the type of brain one has is dependent on individual choice and the power of one's will—mind can override brain.

The rhetoric of the plastic brain is clearly an attractive and influential discourse, one that provides a compelling way to articulate visions of unrestrained optimization at the individual and social levels. Channeling desires for improvement through brain biology gives these desires a sense of clear focus, situating the brain as a visible, calculable space that can be altered through very specific forms of management and intervention. As powerful as these discourses of plasticity are, however, they do not operate in isolation. They work with discourses of determinism to produce an extraordinarily powerful way of accounting for life. It is because determinism—the idea that the brain is the cause of human function—still holds sway that optimization discourses acquire their true strength. The idea of changing the brain would not be so alluring if the brain were not seen as the very source of our humanity—for example, thoughts of a plastic liver or even a plastic heart might be intriguing but are unlikely to generate the same level of excitement and exhilarating sense of possibility as the plastic brain. If we can access our brains, we can intervene at the source and fundamentally alter virtually every aspect of our lives—we can eliminate our individual sadness and collective evil, just as we can construct ourselves as happier, more productive citizens and more efficient, humane societies. Popular neuroscience's ability to oscillate between these different versions of causality—through our will, we cause the brain, and the brain both enables and sometime hijacks our will—is not a problematic contradiction, but rather a source of rhetorical power.

Plasticity and the Politics of Labor

The rhetoric of plasticity supports a conception of humans as adaptable, malleable creatures that can willfully change themselves as necessary to pursue new opportunities or achieve higher levels of their potential. This understanding of human nature has broad currency across a range of contexts, and it influences a whole economic discourse on the individual's role in relation to employment, career, and the workplace. Specifically, there is extraordinary resonance between Begley's plastic brain and recent conversations about work. Just as understandings of the brain are shifting away from ideas including the module or region as the unit of analysis, the permanence of structure-function relationship, and a priori specialization, conceptions of work are following this same pattern. In his recent book *Free Agent Nation: The Future of Working for Yourself*, Daniel Pink describes a shift in the politics of labor: "Large permanent organizations with fixed rosters of individuals are giving way to small, flexible networks with ever-changing collections of talent."[81] If "neuron" is substituted for "individual" in Pink's description, the statement is one that could have been lifted directly from the pages of a contemporary neuroscience text. The points of resonance are multiple: as brain science focuses on smaller units, moving from the region to the neuron, so, too, "the individual, not the organization, has become the economy's fundamental unit."[82] Just as localization hypotheses are giving way to conceptions of mobile neurons, so, too, workers no longer strive for a "fixed place in the world," but instead for mobility that allows them to maximize their talents and productivity.[83] And just as plasticity theories view the neuron's function as fluid and variable, workers are increasingly encouraged to "flexibilize" and adapt themselves to the ever-changing demands of the marketplace.[84]

Thus, while the discourse of phrenology resonated with a particular politics of labor assuming a one-to-one fit between individual and career, Begley's book shares the assumptions and vocabularies of neoliberal economics. The neuron, like the neoliberal worker, is not restrained by spatial location but is free to carry out whatever function is most desirable. This function is not a fixed career, but rather a flexible occupation that can be abandoned or modified at any time to serve the goals of optimization. Neurons, like workers, are required to "behave nimbly, to be open to change on short notice, to take risks continually, to become ever less dependent on regulations or formal procedures."[85] In both the discourse of economics and neuroscience, the politics of flexible labor is framed in terms of choice and freedom. Individuals are not fixed by biology—they are free to improve themselves, virtually without limit, by changing their biology in order to excel in sports, the arts, career, and character. Nor are individuals constrained by their jobs—they are "empowered" instead of "controlled," liberated to "grow and learn" in new ways as they adapt to new institutional demands and corporate expectations.[86] Moreover, flexibility is achieved through a sundering of work from a fixed location (the office, the factory) as work is increasingly expected to take place in non-workplace environments (the home, the car). In both biological and economic configurations, the elements (neurons and workers, respectively) are defined in terms of freedom operationalized as adaptability of function and spatial mobility. This articulation of space, function, and identity stands in striking contrast to the assemblage mapped out by phrenology, a distinction that illustrates the synchronic resonance of science and society across two different historical configurations.

By drawing attention to this resonance between the languages of the brain and the languages of the neoliberal economy, I am not simply pointing to a striking coincidence. Resonance suggests more than shared patterns; it suggests that there is a relationship of mutual influence between brain culture and neoliberal culture. In other words, the ways we understand the brain directly feed into and enable the ways we organize and practice our social, political, and economic worlds. In the following case studies, I trace out these relationships of mutual constitution and enablement to better conceptualize what is at stake in brain culture, and what types of effects brain images produce—effects that reverberate far beyond the spheres of science and medicine. In the next chapter, the first case study takes on the specific question of how individuals are induced to view neuroscience as a practical and empowering discourse that they can use to situate, frame, and regulate their own lives. The broader sociopolitical consequences of this micro-political persuasion are examined in greater depth in subsequent chapters.

3

Practical Neuroscience and Brain-Based Self-Help

Visually confronting one's own brain scan is a transformational encounter fraught with emotion and anticipation, as the neuroscientist and clinical brain imaging expert Daniel Amen vividly dramatizes throughout his dozens of books, television appearances, newspaper columns, and blog entries. In tale after tale, Amen tells the stories of countless patients whose lives have been powerfully changed by the mere act of witnessing their own scan. In a blog entry, "SPECT Scans Offer Hope and Affirmation," Amen quotes one of his physicians, who describes the event of viewing one's scan as an epiphanic experience with religious overtones: "When you show someone there are biological underpinnings of the way that they feel a wave of relief and reassurance washes over them. This affirmation is akin to grace—an undeserved gift. The patient believes they deserve the way they feel because of things that have happened to them. But we can show them our physical body is susceptible to illness and disease and these changes in their brain show our fragility, this mysterious affirmation imparts peace in their very soul."[1] Amen's personal experiences with patients attest to the saving grace of viewing one's scan. For instance, Amen tells the story of Sally, a young woman who suffered from depression and anxiety and had a less than desirable job because she had not finished college. Amen writes, "Seeing the SPECT pictures was very powerful for Sally," as she was able to look and see that her problems were not her "fault." Viewing the scan was akin to a conversion experience, and it "changed her whole perception of herself."[2] After this life-changing encounter, Sally's mood improved and she went back to school to finish college. In another story, Amen tells of Josey, a young woman who suffered from crippling anxiety and inexplicable anger toward God. Viewing her scan was a vital part of her transformation: "Seeing the physiological problem in her brain for herself was the first step in the healing process."[3]

Through her encounter with the visual evidence of her scan, Josey was convinced that her problems were not the result of an immoral or defective character, but rather were medical problems demanding scientifically validated treatments. After her scan-induced transformation, Josey was willing to accept the medical reality of her condition and pursue a treatment program, which included both medication and spiritual meditation exercise. Over a period of months, she was "restored" to health, as the brain treatment program gave her "access to her soul, her real self, and even her God."[4] In story after story, Amen recounts the joys of numerous patients like Sally and Josey who received "grace" and "peace" by seeing their scans with their own eyes, and accepting unconditionally the biological nature of their problems.

Just as the act of viewing one's scan can produce joy, relief, and peace, it can also produce the opposite emotions—fear, dread, and guilt. Strewn among the stories of saving grace are tales in which the encounter with one's scan produces a shock of recognition of one's own corruption and immorality. Amen tells the story of Mark, a young male who was a known drug abuser. Mark's scan showed "multiple holes across the top surface of his brain." When confronted with his scan, Mark was "very upset" and distraught.[5] Far from activating feelings of relief and peace, Mark's visual encounter with his scan resulted in fear and anxiety. In his stories, Amen views these experiences of negative emotion as valuable because they ultimately lead the patient to make different choices. Mark's devastation at seeing his frayed brain brought him to a point where he saw clearly the need to change his lifestyle. Amen has found brain scans to be such a powerful tool for inducing fear and guilt that he designed an award-winning poster displayed in high schools, prisons, and other institutions around the country. The poster, seen in figure 6, features a pattern of brightly colored scans, one full and whole, and the others riddled with holes, graphically suggesting corrosion and decay, headed by the looming text "Which Brain Do You Want?" The poster highlights that brain health is a choice, and that individual will is the deciding factor in whether one's brain will be robust, symmetrical, and whole, or riddled with holes.

How to reconcile these two competing paradigms? On the one hand, viewing one's brain scan is associated with the message that the brain causes negative life events, and therefore that individuals should not hold themselves (their "true selves"—their personalities, souls, or characters) to blame. The brain scan is the conveyor of a message—"Look at what your brain is doing to you!"—that Amen and his patients interpret as one of grace, peace, hope, and affirmation. On the other hand, viewing a brain scan is also associated with the message that personal life choices directly cause the type of brain one has, and therefore that individuals are ultimately responsible for the state of their biological brains. In this case, the scan screams, "Look at what you are doing to your brain!," preaching, in Amen's interpretation, a message of fear, guilt, and intensification of anxiety. What is interesting about popular neuroscience, and Amen's work in particular, is that it does not choose between these competing modes of assigning meaning to brain scans. Scans are illustrative of both individual helplessness in the face of biological factors beyond one's control, and personal obligation to actively manage and direct one's life to optimize brain biology. Whether the scan is taken in an affirming or condemning sense seems to depend entirely on the context in which it is deployed and the meanings attributed to the scan by expert discourses. As Amen's examples show, brain images are flexible rhetorical resources that can support a range of claims about the relationships between identity, will, and the determining powers of biological nature.

In this chapter, I examine Amen's contributions to popular neuroscience, focusing specifically on his recent self-help book *Making a Good Brain Great: The Amen Clinic Program for Achieving and Sustaining Optimal Mental Performance*. My objective is twofold. First, I am interested in the mutually beneficial merger of self-help, a hallmark therapeutic discourse widely recognized as a key contributor to practices of self-governance, and popular neuroscience, with its ubiquitous brain images and elevation of biological vocabularies.[6] Popular neuroscience facilitates self-help's therapeutic agenda by imbuing it with medical and scientific authority backed by visual data. Through the authority of brain images, popular neuroscience helps to bring virtually every aspect of daily life into the domain of the therapeutic, framing everyday practices as amenable to precise scientific assessment and modification through medically authorized treatment regimens. Just as popular neuroscience lends scientific authority to self-help, self-help offers neuroscience productive channels of dissemination through an established form that effectively conditions audiences to actively take up and act on its messages.

My second major objective is to unpack the rhetorical dimensions of health in the context of popular neuroscience. As Amen's discourse reveals, health is an accommodating rhetoric that entwines science, politics, and morality through its polysemous circulation. In other words, as health travels across brain culture's discursive terrains, it is imbued with many meanings—biological, moral, and social. For example, health can be attributed to biological fitness ("he has a healthy brain"), character ("she has a healthy sense of empathy"), and social behavior ("they are involved in a healthy relationship"). By equivocating among and combining these diverse meanings, Amen draws on the languages of health to attribute causal powers alternately to biology or character, often defined in terms of "will." In general, Amen attributes undesirable attitudes and behaviors to brain dysfunction and equates desirable behaviors with brain health. At the same time, brain health is equated with one's "true self," while brain dysfunction is described as usurpation of the self. Through these patterns of attribution, Amen exemplifies a rhetorical process whereby individuals are motivated to act in various ways through appeals to become their real selves. This appeal—"be all you can be," or "be your best self"—is a powerful inducement, freighted with moral connotations, that ties individuals to a number of projects of self-governance. The project of "being your best self" is articulated in the language of health, making it at once a medical project tied to objective scientific knowledge and expertise, and a deeply personal, moral project that emanates from the most intimate depths of one's being. Because health is able to trigger both scientific and moral meanings, it is a powerful language for communicating self-governance as an intimately personal and simultaneously objectively scientific affair.

In following sections, I describe Amen's prominent role in brain-based self-help discourses, and then examine the rhetorical effects of the fusion of self-help and popular neuroscience. The remainder of the chapter focuses on Amen's recent brain-based self-help book *Making a Good Brain Great*, and works to identify the ways in which brain scans and vocabularies of brain health are deployed to frame life as an ongoing project of self-realization, one that requires sustained treatment programs comprising biological, psychological, and spiritual remedies.

Amen's Brain-Based Self-Help Empire

In his 2007 book, provocatively titled *Sex on the Brain: 12 Lessons to Enhance Your Love Life*, Amen includes a chapter called "Look Closely: Brain Imaging Secrets to Enhance Your Love Life." In this chapter, Amen explains that if people only pay attention, the truths revealed by brain imaging will lead to great sex, amazing relationships, and incredible fulfillment. Whether people are struggling with sexual difficulties, communication problems, or even divorce, the clear answer in each case is "targeted brain help."[7] When scans show brain activity that is unhealthy, Amen often prescribes medications, producing what he calls "better marriage through biochemistry."[8] In addition, he suggests a variety of other means of improving one's brain function, the ultimate way to achieve great sex and healthy relationships. These therapies include diet tips, sleep, and various forms of mental exercise. By conceptualizing the brain as something that can be seen, managed, and measured, people can take charge of their lives, acting on the brain in order to act on their lives. Through self-management of the brain, it is possible for everyone to achieve amazing sex and relational satisfaction.

Sex on the Brain is just one small part of Amen's extraordinary corpus of brain rhetoric. The clinical neuroscientist, psychiatrist, and brain-imaging expert is in many ways a central node in his own rhetorical or discursive formation. Amen is the head of the four Amen Clinics, all of which specialize in the clinical use of SPECT (single photon emission computed tomography) brain imaging. He has a significant presence in the public sphere as the author of twenty books (translated into thirteen languages), a number of audio and video programs, a regular column in *Men's Health*, and numerous articles and interviews in popular sources such as *Newsweek*. Amen regularly appears on popular television shows, including *Today*, *Ricki Lake*, *The View*, and various CNN programs, as well as radio shows and speaking engagements. He has won awards for his role in creating antidrug campaigns, as well as other areas of his research and writing. His book *Change Your Brain, Change Your Life* was a *New York Times* best seller in 1999, unexpectedly selling tens of thousands of copies in its first year, in part because, according to *Publishers Weekly*, it "struck a nerve with readers who love a 'scientific' hook."[9]

The majority of Amen's work, including his recent *Making a Good Brain Great*, follows the pattern of *Sex on the Brain*—different areas of life (education, family, fitness, employment) are translated into brain issues through the use of imaging technologies and biological vocabularies. After transforming diverse life areas into brain-focused issues, Amen provides concrete recommendations for improvement. This advice is relatively consistent across the different areas—time and again, Amen exhorts healthy diets, quality sleep, and meditative exercises designed to cultivate focus and relaxation. Amen is also an outspoken advocate of psychotropic medications, and his books and blog are riddled with references to name-brand drugs heralded for working wonders in correcting specific maladies linked to brain and character malfunction.

Although Amen claims an association with the prestigious UC Irvine's School of Medicine and the American Psychiatric Association, he is not unconditionally accepted as a legitimate expert.[10] Amen's broad clinical applications of SPECT brain scans are not widely accepted by the neuroscientific community. The scientific consensus is that brain images have great potential as a research tool but limited practical applications in clinical settings. Notably, in Amen's book, brain images are not used as a tool to diagnose illness or inadequacy; rather, they are used as inventional resources to articulate neuroscience as a practical aid to everyday living. The brain, as visualized in a scan, becomes a discursive space for coordinating the personal, social, and biological dimensions of life and translating them into the common language of health. In the remainder of this chapter, I examine how Amen uses brain images to disseminate biological vocabularies and ways of acting to audiences, correlating ethical citizenship and personal fulfillment through brain images that translate both into calculable forms amenable to technical intervention in the name of health.

Practical Neuroscience and Self-Help

Sex on the Brain illustrates two important facets of the rhetorical brain. First, Amen's popular book shows how brain imaging can be used to turn all aspects of everyday living, from the specific (diet) to the abstract (aspirations and moods), into a concrete "object" (the brain) that can be regulated, assessed, and manipulated. Caring for the brain becomes the means of achieving abstract goals such as fulfillment, happiness, and, in this case, definitive sexual gratification. Brain imaging is an important part of this message because it visualizes life in a way that is accessible to science, correlating all the habits, aspirations, and relationships that make up daily living with biological events that can be seen in the form of a digital image. Second, *Sex on the Brain* shows the extent to which popular media (including scientific messages directed to public audiences) articulate contemporary neuroscience as a set of guidelines that can be applied to everyday life. Amen lauds this as "practical neuroscience," a tool for daily living that is useful for everyone, not just experts and scientists.

One avenue of dissemination of practical neuroscience is popular literature, including Steven Johnson's *Mind Wide Open: Neuroscience and Everyday Life* and Sharon Begley's *Train Your Mind, Change Your Brain*. These books, generally written by nonscientists, summarize and seek to utilize the latest neuroscientific research. In other instances, neuroscientists address public audiences directly, framing their own research as relevant to everyday life. One prominent form that neuroscientists use to address public audiences about the practical benefits of brain science is the instructional handbook that frames neuroscience as a comprehensive toolbox of techniques for living. Recent examples of these handbooks include Jeffrey Schwartz's *Brain Lock: A Four-Step Self-Treatment Method to Change Your Brain Chemistry*; Richard Restak's *Mozart's Brain and the Fighter Pilot: Unleashing Your Brain's Potential*; and Daniel Amen's many books, including the one I focus on in this chapter, *Making a Good Brain Great: The Amen Clinic Program for Achieving and Sustaining Optimal Mental Performance*.

In general, self-help books are perhaps the hallmark discursive genre of contemporary culture. The market for self-help books has exploded in the last decade, with the constant demand leading *Publishers Weekly* to dub the genre a "Teflon category."[11] The market for self-improvement literature grew 50 percent between 2000 and 2004 and is currently an $8.56 billion annual business.[12] Statistics suggest that one-third to one-half of all Americans have purchased a self-help book at some point in their lives, making the genre a powerful economic and cultural force.[13] One of the primary functions of self-help books is to make various dimensions of life—career, relationships, spirituality—manageable or amenable to technical intervention. In other words, they articulate life as something calculable that can be intention-ally managed to achieve desired outcomes. Self-help promises individuals that they can reflexively act on their own lives, becoming experts of the self who know exactly how to manage themselves—including how to optimize their relationships, fulfill their spiritual inclinations, and achieve success in their careers.

Sandra Dolby has defined self-help as a specific form of literature with four generic characteristics.[14] First, the content of self-help books is nonfic-tion and aimed toward self-improvement. Second, the books are written in a relatively informal style, with the author ("I") addressing the reader ("you") in a direct manner. In other words, the books are widely accessible and not aimed to an audience of experts who must possess a specialized vocabulary. Third, the books all take the form of a problem-solution discourse—they identify a problem, or problem area, and then offer practical ways to man-age, overcome, or otherwise deal with the problem. Finally, the books are designed to educate readers—readers must become the active ones who solve their own problems. This defines the self-help philosophy and sets the genre apart: individuals cannot simply take the advice of experts to achieve genuine transformation; rather, their cure depends on their own acquisition of knowledge.

The generic characteristics of self-help literature make it a compelling way to translate neuroscience into practical knowledge. Because the purpose of self-help is to get readers to frame and manage their own lives in particu-lar ways, or to become experts of themselves, in the context of neuroscience it works to distribute brain-based modes of understanding and acting on the self. Brain-based self-help books are written by neuroscience experts (neuroscientists or neuropsychiatrists), and the biology of the brain is cen-tral to their content. Moreover, brain-based self-help features brain imaging as a central persuasive resource. Amen's *Making a Good Brain Great* features a number of scan images, which are used to illustrate how brain biology con-tributes to problems as well as to provide proof of how self-help techniques can alter this biology in positive ways. Amen's book shows how brain imag-ing is far more than a scientific research agenda—it is a persuasive visual rhetoric by which neuroscience is articulated as relevant to the construction and maintenance of desirable selves.

In the context of brain-based self-help, brain images contribute to the significance of health as a governing value and promote what Paul Rabi-now describes as the "working on oneself in a continuous fashion so as to produce an efficient and adaptable subject."[15] Self-help's deployments of brain images make it appear as if specific regimens (diet, exercise, and so on) modify the brain in very specific ways, ultimately resulting in an improved life. These desirable outcomes—improved life, fulfillment, happiness, success—are described in terms of health. Health is not simply a biomedical concept, but a general telos of self-improvement.[16] As Rabinow suggests, the implications of this continuous self-fashioning in the name of health are not limited to the personal realm; they also have political consequences. Popularized neuroscience discourses encourage particular *kinds* of selves, or particular types of citizens, who are more or less amenable to diverse political agendas. My attention to the micro-political processes of "making up" citizens prefaces the later case studies in this book that take on the macro-political consequences of these transformations in public discourse. In this chapter, my focus is limited to an analysis of Amen's text as a representative example of brain-based self-help and, more broadly, popularizations of neuroscience in order to analyze the micro-rhetorical operation of the growing diffusion of neuroscience across the public grammar.

Fifteen Days to a Better Brain!

Amen's *Making a Good Brain Great* is constructed as an educational "how-to" manual, promising readers that they can improve their brains in as little as fifteen days. The book is divided into two sections. The first details nine "brain-centered principles to change your life," with each chapter devoted to a specific principle. The primary purpose of these chapters is educational. The main theme is that the brain is a major actor in virtually all areas of life. The first principle encapsulates this theme: "Your brain is involved in everything you do." Throughout the first half of the book, Amen instructs his readers on the biology of the brain, the benefits of imaging technologies, and the different categories of living that are determined by brain biology. In the second half, Amen presents "The Amen Clinic Program for Making a Good Brain Great." This is the "how-to" portion of the book. Each chapter is dedicated to a different set of tasks: maintaining a healthy diet, exercising regularly, listening to soothing music, managing stress, and having positive social interactions, including regular sexual relations. In the concluding chapter, Amen combines these tasks into a fifteen-day program, with each day devoted to improvement in a specific area. For example, day 6 is "Great Music Day," and day 9 is "Grateful Day." Every day includes imperatives to exercise, meditate, and meticulously track one's diet.

In terms of the content of his suggestions, Amen hardly breaks new ground. In many respects, the second half of the book reads like an annotated amalgam of existing self-help books, each devoted to a specific topic such as diet, exercise, social interaction, or stress management. For instance,

in the diet chapter Amen condemns those foods that are "laden with calo-
ries, refined carbohydrates, and damaged fats," indicting the fast-food culture that thrives on "super-sizing" unhealthy products.[17] His alternative is
to increase water intake while decreasing consumption of the bad foods
and focusing on a diet of protein, good fats, and carbohydrates. The chapter
even includes a collection of healthy recipes, ranging from a low-fat chicken
omelet to blueberry ice cream. There is nothing novel about these recommendations, and they are so commonplace that they verge on the clichéd.
The difference is that these are described as recommendations designed not
for the health of the person but the health of the brain. Food is a "powerful
brain medicine" and the suggestions are "brain-promoting nutritional tips"
(*MGBG*, 89, 91). Calorie restriction is "helpful for the brain," and the recipes
are "brain-healthy" (*MGBG*, 89, 104).

The rest of the chapters follow this pattern. Commonplace recommendations for self-improvement are reiterated, this time in terms of their
effects on the brain. Physical exercise is important because it allows the
brain to generate new neurons, and coordination activities are lauded for
their brain-enhancement potentials. Regular sexual relations are vital to a
healthy brain, because it is "the largest sex organ in the body," involved in
everything one does, "including everything sexual" (*MGBG*, 134). "Weird"
sexual fetishes are "brain symptoms" (*MGBG*, 143). The how-to recommendations are all consistent with this formula: take existing, commonsensical
knowledge (including social mores) and rearticulate it in terms of how it
relates to the brain as both cause and consequence of behavior.

Amen brings all of life, down to the most mundane details (e.g., individuals are advised to use real, rather than artificially flavored, vanilla
extract in the milk they are to consume before bedtime to ensure proper
rest), within the domain of neuroscience. Every aspect of life presents a
potential opportunity for self-fashioning and improvement. The ubiquitous
brain images populating the book's pages provide support for this assimilation of life into the rubric of what Amen calls practical neuroscience.
Because brain images can be produced in correlation with virtually any
activity or differentiated identity (specific thoughts, emotions, behaviors,
and specific types of persons, such as depressed, female, or young), they can
provide visual evidence that all of these activities and identities are biological and, more specifically, neurological. Further, because the brain images
are dynamic and can visually depict neurological changes in coordination
with different technical interventions (including medication, behavior
therapy, and "positive thinking"), diverse medical, social, and personal
technologies are articulated in terms of their value for health. In Amen's
book, brain images are both an essential part of the verbal narrative and
an important material feature of the text. Amen frequently references brain

images by calling on descriptions of the imaging process and the images themselves to support his claims, and he includes multiple representations of brain images for the readers' own consumption. In the next section, I describe the types of images Amen uses, with specific attention paid to their visual composition.

"How Do You *Know* Unless You *Look*?"

This is the title of Amen's seventh chapter (emphasis mine), and it indicates that knowledge is intimately connected to, and in fact dependent on, looking. Brain images are central to Amen's project and they are the evidence that grounds both the nine brain-centered principles and the detailed recommendations for brain improvement. Amen is recognized for using SPECT imaging in clinical settings. His clinics have performed more than fifty thousand of these scans, constituting the largest database of SPECT images in the world. SPECT imaging is a nuclear medicine procedure that measures brain blood flow. The assumption is that brain blood flow is correlated with brain activity. The data is constructed into three-dimensional images that model these patterns of brain activity. Amen's books are replete with black-and-white representations of these images, and his website and educational pamphlets include colored representations of SPECT images.

The SPECT images are presented as visual evidence that is highly legible to even an untrained audience. Amen explains, "SPECT scans look at function or how the brain works. SPECT results are actually very easy to read and understand. We look at areas of the brain that work well, areas that work too hard, and areas that do not work hard enough" (*MGBG*, 8). Later he writes, "Scans must be clear, understandable, easily illustrative of brain function, and available to the patient on a timely basis. We believe our three-dimensional rendering software makes the scans easy for professionals, parents, and families to understand" (*MGBG*, 255). Throughout *Making a Good Brain Great*, SPECT scans are presented as legible through Amen's description of their usage in clinical settings, and in the way the actual images are framed and presented.

There are two types of SPECT images: three-dimensional surface images and three-dimensional active images. The surface images depict blood flow at the brain's cortical surface and are set to display the top 45 percent of brain activity. These images display brain function or activity, but they look like "objects," or representative models of brain structure. They have an apparent density and solidity, which fosters the impression that if the referent were present, one could pick it up and hold it for visual and tactile inspection. The three-dimensional surface image corresponding to a healthy brain shows "full, symmetrical activity," and looks like a clay model

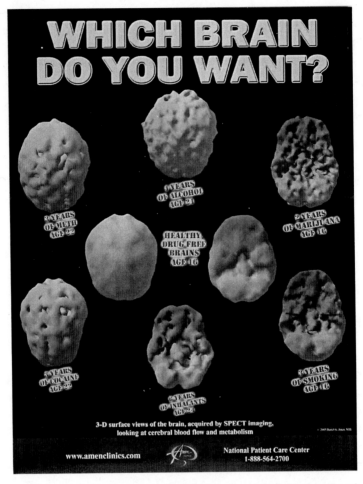

FIGURE 6 SPECT images from one of the Amen Clinic's educational posters. The display of images shows how Amen uses SPECT to illustrate "before" and "after" states of the brain, visibly illustrating the effects of various stimuli. In this poster, the effects of drug use are graphically depicted in such a way as to suggest that the consumption of narcotics literally eats away at brain structures.

Poster image courtesy of Amen Clinics, Inc., used with permission

of an actual brain. It appears to be smooth, and there are no holes or gaps in the surface. The images of unhealthy brains appear to be moth-eaten. They show dramatic holes and gaps rather than a smooth structure, and they suggest that the brain physically decays when it is correlated with the unhealthy state (due to such factors as negative thinking or drug use). In the book, all the images are in black-white or grayscale. The colored images featured on the poster and on Amen's websites are even more dramatic. The brains are depicted in varying shades of yellow, orange, purple, and blue. The top-down view of the healthy brain is an organized blend of yellows and reds, each shade subtly blending into the next. The unhealthy brains, covered in holes and structural defects, display sharply contrasting shades to illustrate the "underactivity" or "overactivity" associated with a particular area.

The other images are three-dimensional active images, which compare average brain activity to the "hottest" 15 percent of activity. These images look like graphs: the whole brain is modeled as a three-dimensional grid displaying fine lines interconnected in web-like fashion. Specific parts of this three-dimensional diagram are filled in with shading to suggest overactivity. Although these images appear more like graphs than the surface images, which look like actual representations, they suggest a simple relationship between brain activity and the image. The complex averaging procedures and statistical work that goes into producing these images are lost in the neat, simple-looking visuals that are presented for the readers' consumption and interpretation.

Amen's three-dimensional active and three-dimensional surface images support a blended localization-connectionist discourse. Amen largely talks of the brain as a space of connections and interconnections, stating that it is important to understand that brain systems do not exist in a vacuum, but rather "are intricately interconnected," such that "whenever you affect one system, you're likely to affect the others as well."[18] Brain problems are not the result of anatomical or structural issues, Amen emphasizes, but instead are problems of function. Restoring brain health is often talked about in connectionist vocabularies, in terms of rebalancing activity, increasing connectivity, smoothing out function, and correcting for too much or too little activity. At the same time, Amen associates distinctive behavior patterns with particular systems that occupy identifiable spaces in the brain. Through this framework, Amen effectively frames brain health as both a tenuous process of balance and connectivity, and something that can be modified through focused interventions that target specific brain systems. Both types of images Amen uses suggest regional specificity, and the three-dimensional active images in particular constitute the brain as a space of interconnected, distributed networks with fluctuating intensities of activity.

Although I am not primarily concerned with the "truth-value" of these images, it is worth pointing out that, like all functional brain imaging, there is no scientific consensus as to their diagnostic value, and most medical authorities tend to be skeptical of Amen's broad clinical applications.[19] In other words, scientific consensus tends to hold that it is not possible to

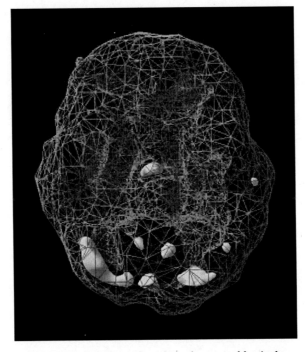

FIGURE 7 Underside, active view of a normal brain from
Amen's SPECT atlas.

Reproduced from Daniel Amen, *Images into Human Behavior:*
A Brain SPECT Atlas. Image used with permission, courtesy
Amen Clinics, Inc.

determine from a brain image whether that person is healthy or whether
they have a particular disorder. This is in part because brain activity varies
substantially from individual to individual, and most research studies average data from many individuals to produce results about specific populations. Amen explicitly recognizes his divergence from mainstream medical
views, and he frequently defines himself as a "heretic" who draws unjust
criticism for his courage and foresight. In his discussion of imaging, however, Amen avoids claiming that he can make a diagnosis on the basis of a
scan alone. As I will explain, Amen defines a healthy brain by way of external
behavior: a healthy brain is identified, in other words, if the individual is
a good citizen. This circularity highlights the fact that the images are not
primarily diagnostic tools, but rather persuasive tools that have functions
independent of their medical utility. They impel individuals to frame their
lives within the context of neuroscientific knowledge, thinking of and acting
on their lives, motivated by the never-ending pursuit of health.

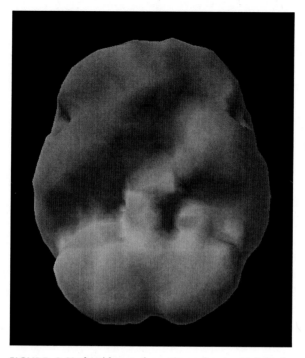

FIGURE 8 Underside, surface view of a normal brain from Amen's SPECT atlas.

Reproduced from Daniel Amen, *Images into Human Behavior: A Brain SPECT Atlas*. Image used with permission, courtesy Amen Clinics, Inc.

Practical Neuroscience as a Program for Living

Nikolas Rose writes that the dispersal of biological, including neuroscientific, vocabularies "make[s] up citizens in new ways."[20] In other words, as individuals come to take up the biomedical terminologies disseminated by experts and adopt them as vocabularies capable of expressing their own problems and aspirations, they speak and act differently. As Rose explains, "They use these phrases, and the types of calculation to which they are attached, to make judgments as to how they could or should act, the kinds of things they fear, and the kinds of lives for which they can hope."[21] This dissemination of biomedical language exhibits two important features of neuroscience as a mode of government: First, terms and ideas must not only be transmitted, but also actively adopted by their audience. Second, the terms are connected to ways of acting. If the vocabulary is adopted, specific modes of calculation and intervention will follow. Amen's book illustrates this complex form of dissemination. As medical advice, Amen's address is distinguished by its insistence that the patient *understand* his or her own problems. More important, individuals must understand their problems through the biological vocabulary of neuroscience. It is not enough for individuals to accept Amen's prescriptions because of his expertise and authority; each person must come to a scientific understanding of his or her life experiences. Brain images authorize neuroscience as a "truthful discourse," the correct and ethical way of accounting for one's self and others.

In the brain-based self-help books, knowledge is intimately tied to health. Readers are induced to adopt a neuroscientific vocabulary as a necessary component of the self-improvement program. Amen's fifth principle, for instance, is "Know and heal the brain systems that run your life." This chapter is intended to educate readers about the different functions of six brain systems. But this understanding is not merely educational; it is a vital component of transformational healing. Amen begins the chapter thus: "To make a good brain great, it is important to have a basic understanding of how the brain works, including its strengths and weaknesses" (*MGBG*, 32). Amen presents a "hands-on guide" to understanding the brain, emphasizing that it is essential to know "that certain parts of the brain tend to do certain things, and that problems in specific areas tend to cause identifiable troubles" (*MGBG*, 32). When one understands the way the brain functions, it is then possible to identify the precise location of problems and begin "targeted treatment."

This hands-on guide differs from impersonal pedagogical instruction in that it is highly participatory. Readers are invited to learn not about *the* brain, but about *their* brain. Amen presents the Amen Clinic Brain System Quiz, a self-report designed to assess the function of brain biology. The quiz asks readers to rate themselves on sixty different symptoms from zero (never) to four (very frequently). The quiz includes items covering cognitive, emotional, and physical health, such as "difficulty expressing empathy for others," "upset when things do not go your way," "trouble learning new information," and "fingernail biting." Each question is correlated with a specific area of the brain—prefrontal cortex, basal ganglia, and so forth. Individuals associate their answers with the appropriate brain area to assess overactivity or underactivity in each region. For example, inconsistency, troubled decisions, and impulsive, risky behavior (such as sexual promiscuity or drug abuse) are interpreted as signs of pathology in the function of the prefrontal cortex (*MGBG*, 39). Worrying excessively should be understood as "high activity in the ACG," or the anterior cingulate gyrus. Each brain system is accompanied by a full-page table that lists the types of problems associated with underactivity or overactivity in that particular region.

The quiz conditions readers to understand their subjective experiences in biological terminologies that emphasize their neural origins. Readers are instructed in neuroscientific vocabularies and, more generally, in a logic of correlation that articulates personal and social events as fundamentally biological. Neuroscience is held out as a language that can be used to express everyday experiences, as in "it's a low serotonin day," or "her PFC [prefrontal cortex] is clearly overactive." These vocabularies are taken up as ways of understanding not only the self, but also the behaviors of others. In *Change Your Brain, Change Your Life*, for instance, Amen writes that children with OCD often have "cingulate parents."[22] Readers are advised that if they are engaging in an argument with someone who clearly has a "cingulate" problem, the best thing to do is avoid the conflict by retreating to the bathroom with a large book.[23] Neuroscience becomes a colloquial language to understand social behaviors. Here, cingulate is used as an adjective that describes not biology but observed character. Although the visual scan is used to provide evidence that neuroscience is a truthful discourse for self-expression, the image is not a continual necessity for the attribution of observed behaviors to events in the brain. In other words, the brain images enable the assimilation of personal and social events into a neuroscientific paradigm by showing their biological nature, but once this biosocial paradigm is established, it enables a direct movement from the biological to the personal or social without an actual mediation of the image in each instance. Amen's articulation of his quiz illustrates this: he emphasizes that readers do not actually need to get a scan to benefit from his books, as simply taking the quizzes and becoming fluent in the languages of brain biology are all it takes to successfully participate in a targeted brain treatment program. The role of brain images, then, is not primarily diagnostic, but rather rhetorical—to facilitate the dissemination and adoption of brain-based vocabularies and paradigms for living.

The neurobiological vocabulary is not simply one among many: it is *the* way of understanding experience that one *should* adopt, to the exclusion of all alternatives. The brain-based self-help book compels readers to adopt this terminology through both scientific and moral appeals. First, by virtue of his scientific authority, bolstered by the presence of the images, Amen situates neuroscience as a valid and truthful vocabulary. He claims that he has looked at more brain scan images than any other person. Because the SPECT images are framed as accessible to his readers, they can share this expertise and extend it across their own lives.

In addition, the neuroscientific paradigm is the morally correct way to understand the behaviors and motives of both the self and others. Vocabularies of morality are coordinated with the value of health. A failure to adopt the neuroscientific paradigm results in "stigma" and "discrimination," attitudinal barriers to healing and a sign of pathological social behavior. Understanding personal and social events in biological terms is itself a type of healing because it eliminates destructive attitudes such as guilt and self-blame. Alternative accounts are framed as "prejudice" or harmful ignorance. Amen writes that a biological understanding results in "reduced shame, guilt, stigma, and self-loathing. This understanding can promote self-forgiveness, often the first step in healing" (*MGBG*, 252). The stories of Sally and Josey exemplify this theme, that negative thoughts produced by self-blaming are themselves causes of pathological brain patterns. Accepting that one's behavior is caused by biological factors outside one's control is both a type of self-knowledge and a material action that literally changes the self.

This moral imperative to adopt a biological framework applies to the behaviors of others as well, including behaviors with significant social consequences. For instance, Amen has testified in court numerous times, using the evidence of brain scans to argue that criminal acts should be treated as illnesses rather than acts of social deviance. For example, he talks about a man who was upset because he thought that his wife was neglecting him. When he became convinced by brain scans that her behavior was biological in origin, he no longer blamed his wife for harboring negative feelings toward him. In fact, the husband's attitude of blame toward his wife is in itself a pathological symptom—"stigma"—that restricts his ability to achieve optimal health.

Are You Your Brain?: Identifying and Disidentifying Person and Scan

Amen's brain-based self-help rhetoric relies on strategies of both identification and disidentification to encourage viewers to adopt brain-optimizing treatment programs. Strategies of identification invite patients to understand their scans as images of their own selves, visualizations that access and display their biological as well as psychic interiors. These strategies are designed to bring about a "That's me!" recognition upon viewing the scan. As I will explain, the practice of identifying patient with scan encourages individuals to see the scans as relatively straightforward depictions of their own selves, and consequently to act on their selves through brain-based vocabularies and interventions. Yet, at the same time, brain-based self-help utilizes powerful strategies of disidentification, or rhetorical devices that encourage patients to view their scan as "not me," an image of something that is somehow exterior to and distinctive from the self. These strategies of disidentification work in combination with strategies of identification to construct a dividing line between one's current self and one's true self. The true self, in this context, is correlated with a healthy brain and framed as something that must be achieved through medical treatment programs. Images of unhealthy brains reveal the current self in its incomplete, inadequate state, but they also promise a glimpse of the contours of the healthy brain. Thus the scans work on two levels: to reveal the problematic state of the current self, and to point toward a future wholeness, where the scan of a perfectly healthy brain verifies the achievement of one's true nature.

Identification: "You Are Your Brain"

Making a Good Brain Great exemplifies what Joseph Dumit describes as "the cultural and visual logics by which [brain scan] images persuade viewers to equate person with brain, brain with scan."[24] Through these cultural and visual logics, individual character is collapsed into biology by way of brain scans that provide concrete visual evidence of the neurobiological nature of virtually every subjective experience or social interaction. The Amen Clinic Brain System Quiz described above provides an example of this identification. The quiz is framed as a substitute for a brain scan, and the implication is that brain scans provide the same type of knowledge as introspective techniques such as the self-report data. Brain scans can, in this articulation, literally visualize subjective character.

Amen often frames the images in equivocal vocabularies that situate the scans as representations of both biology and character. Looking at a brain scan is equivalent to looking at the ultimate depths of one's true self. In fact, Amen posits a direct relationship between character and biology in his second principle: "When your brain works right, you work right; when your brain is troubled, it is hard to be your best self" (*MGBG*, 7). This equation is authorized by images: "This principle came to me after *looking* at hundreds of scans on my own patients. Not only do I read scans, I also work directly with patients and families, *looking* into the lives of the people behind the images" (*MGBG*, 7; emphasis mine). Two different senses of looking are used here: a literal looking at material images, and a figurative looking into people's lives. Throughout his manual, Amen uses "looking" in an equivocal fashion, identifying the viewing of a scan with understanding and evaluating lives: "Look at any aspect of behavior—relationships, school, work, religion, sports—and in the middle of it all is brain function" (*MGBG*, 4). There is a similar equivocation in the adjectives used to describe both the scan and subjective existence. Amen identifies "the *quality* of brain function represented by the scans" with "the *quality* of the decisions, outcomes, and emotional connections in the lives of my patients" (*MGBG*, 8; emphasis mine). In the first usage, quality can be assessed visually and calculated, and in the second usage, quality is an abstract designation, as in "quality of life." Quality of image is assessed in terms of "underactivity" and "overactivity." The three-dimensional surface images reveal underactivity by depicting certain areas of the brain as gaping holes or moth-eaten structures. The three-dimensional active images mark the overactive areas as "hot spots" on otherwise bare grids of lines. The images are graphical renderings of statistical data, so in the sense of the image, "overactivity" is a quantitative, calculated assessment of brain blood-flow patterns. These terms can also be applied to subjective and social aspects of existence as qualitative judgments about mannerisms or emotions. For example, low activity in the prefrontal cortex is associated with "lack of empathy" and "poor judgment." High activity in the PFC is linked to being "overfocused," "rigid," and "inflexible" (*MGBG*, 38).

"Overactive" and "underactive" evaluate in terms of degree, and these judgments can be either quantitative or qualitative depending on whether a visual rendering of graphic data or subjective experiences and social interactions are being assessed. By fusing these two dimensions—the biological and the personal/social—Amen's expert discourse implies that the personal and social dimensions of living share the qualities of the scan itself and can be calculated, measured, and corrected through precisely calibrated technical interventions. As I will explain, even the subjective aspects of personal life are amenable to quantitative assessment. Because health can be assessed in degrees, access to one's true self (and corollary notions of happiness, quality of life, and fulfillment) can also be assigned quantitative evaluations.

Disidentification: You Are Not Your (Unhealthy) Brain

The identification of brain scan and character occurs against the backdrop of an even more powerful disidentification. At the beginning of this chapter, two sets of examples were described: the first of patients who were advised to see their scans as proof of the causal role biology played in their negative life events ("Look at what your brain is doing to you!"); and the second of patients who were shown scans as evidence of their poor choices ("Look at what you are doing to your brain!"). In both these examples, there is a fundamental disidentification between person and brain, as depicted by the scan. In the first case, the scan visualizes the brain as an actor that affects the self but cannot be equated with the self. This distance is frequently articulated in terms of desire: brain biology hijacks personal volition, literally preventing individuals from being who they want to be and doing what they want to do. For example, Amen quotes a patient who was able to act on his desires only after his brain dysfunction was corrected: "I always wanted to be polite, but my brain wouldn't let me."[25]

In the second set of examples, the brain is objectified as something that bears the effects of personal choices but is not precisely equivalent to the choosing person. These disidentifications, or rhetorical patterns of creating a distance between person and scan, work together with the identifications, which urge people to see their scans as equivalent to their lives. The ultimate message of these complex equivalences and differences is that the actual or current self is revealed in the scan, but because the scan almost always shows a dysfunctional brain, it reveals a flawed, imperfect, and incomplete self. One's true self can correspond only to a perfectly healthy brain, and because almost no one has a perfectly healthy brain, both health and the true self are projects that must be achieved through treatment programs. Moreover, health and the true self are one and the same: the true self manifests itself only when brain function is perfectly and completely optimized.

Amen's distinction between "will-driven" and "brain-driven" behavior further illustrates how the language of health is used to craft a division between the actual self and the real self. For Amen, will-driven behaviors are those chosen by a perfectly functioning, healthy brain. He explains, "Will-driven behavior comes from a healthy brain. It allows you to exert conscious choice over a situation to work in your own best interest" (MGBG, 8). Healthy, will-driven behavior is productive and includes social relationships as well as individual behaviors and attitudes. People with healthy brains "tend to make the best employees, the best husbands and wives, the best parents, friends, employees, and citizens" (MGBG, 11–12). Brain-driven behavior, on the other hand, occurs when the brain does not function in a healthy manner and overpowers the will. Brain dysfunctions deprive individuals of their free will and deny them "access to their true selves" (MGBG, 12). Amen frequently uses the term "hijack" to describe the brain's action on the self. As Amen describes his conclusions of clinical practice, "The brain function of my patients who did bad things was much worse than that of people who were living productive, healthy lives" (MGBG, 15). He summarizes this in the form of a principle: "When your brain works right, you work right, and when the brain is troubled, it is very hard to be your best self!" (MGBG, 16).

Amen's distinction between will-driven and brain-driven behavior shows how health is framed as an intimate, personal, and moral, as much as a biological, pursuit. Health is akin to human "goodness," viewed in terms of social behaviors. Healthy humans are good citizens: in their natural, healthy state, humans are fundamentally good creatures who form positive relationships, excel in the workplace, exhibit attitudes of kindness and compassion, and refrain from criminal or other socially devalued behavior. Any deviation from this ideal form of citizenship is evidence of a brain pathology that is outside the individual's direct control. This goodness is not, however, an abstract moral dictate, but rather a personal value emanating from the deepest, most authentic reaches of one's own best self. Individuals are assured that their healthy, true selves are good, moral, effective, and happy, and that all they have to do is correct their errant biology to finally realize the self they are destined to enjoy.

In Amen's construction, being one's true self is not an all-or-nothing proposition. Rather, the true self is a ground of being one has access to in differing degrees or amounts. Free will, Amen explains, is a quantity that varies. For a completely healthy person, access to this will is 100 percent, yet for a person with brain disorders, such access might be lower, depending on the severity of the disorder. Amen writes, "A patient with OCD or TS [Tourette's syndrome] or drug addiction has significantly less" (*MGBG*, 15). One might have only 80 percent access to one's will, equivalent to 80 percent access to one's true self. Convicted murderers, Amen writes, have considerably lower access to will, leading Amen to oppose capital punishment because, he explains, "killing people with bad brains is akin to killing sick people" (*MGBG*, 15). Thus, Amen concludes, not all people have equal access to their free will, or true selves, and not all should be judged according to the same standards, because different distributions of brain health create a slanted playing field.

The quantification of free will and access to authentic selfhood further amplifies the message that life, including its most subjective, personal dimensions, is calculable and can be assessed in the languages of the brain. Further, this articulation hosts a philosophy of human nature buried in the language of medical science. Individual desires and choices are, by definition, morally correct, rational, and socially acceptable. Any decision that deviates from this normative order is, by definition, not "freely chosen," but rather an effect of biological malfunction. It is impossible, in other words, for bad things to be freely chosen. While this pattern of attributing responsibility for behaviors and attitudes is portrayed as freeing—because it enables individuals to understand that they are not responsible for bad behaviors or deserving of negative events—the alleviation of blame comes with its own costs. The obligation to become healthy and, at the same time, authentically one's self is an intensive, all-encompassing duty that carries its own brand of guilt, baggage, and pressures. Moreover, as Dana Cloud has argued eloquently, these types of therapeutic rhetorics individualize problems and challenge social critique.[26] The brain-based version of therapy discourse is perhaps more pernicious in this regard, because it both individualizes social problems (all social ills can be attributed to brain dysfunction) and articulates social norms as natural, biologically and scientifically true. In Amen's frame, there is no space for individuals to actively choose to violate social codes in order to protest convention or simply out of sheer maliciousness. All such violations are, by definition, pathological. In addition to the consequences associated with therapy's individualizing tendencies, the fusion of health and personal desire also works to channel individuals into certain practices of self-governance, such as those prescribed in Amen's treatment program for achieving health, genuine selfhood, happiness, and success, to which I turn in the next section.

The Pursuit of Health

The ostensibly neutral, scientific value of health provides the framework for the individual pursuit of self-improvement. Because brain images assimilate virtually all areas of life into a neuroscientific rubric, health encompasses behaviors, attitudes, and relationships. As a meta-value regulating existence, health is not a clearly demarcated state that one can reach, but an ever-receding goal that necessitates a continuous pursuit. Health, in other words, can always be improved but can never be achieved. Neuroscientific vocabularies authorize a host of self-fashioning modalities, or "treatment programs," designed to maximize health. Because these technologies are treatment programs, they are simply technical interventions designed to achieve or restore a natural ideal—any political or ideological cast is subordinated. Amen's treatment programs encompass what Mariana Valverde describes as "hybrid" recommendations, which blend moral and physical categories.[27] Treatment programs include, for instance, physical remedies like medication as well as spiritual remedies like meditation designed to act on patients' "souls." In Amen's discourse, both conventionally physical as well as spiritual activities are subsumed under the category of health and articulated with reference to biomedical categories.

One major form of treatment is psychotropic medications. Amen recognizes the persuasive force of the brain scan images when he writes, "A SPECT scan allows patients to see a physical representation of their problems that is accurate and reliable, and that helps to increase compliance—pictures are powerful. It can influence a patient's willingness and ability to accept and adhere to the treatment program" (*MGBG*, 52). The treatment program might include medication but is by no means limited to drug therapy. Amen tells the story of one woman who was "desperate to function as the good mother she wanted to be to her child." Amen diagnosed her with depression and prescribed Prozac. The woman, however, "did not want to see herself in that light or be stigmatized" by the label of mental illness. She stopped taking the medication until Amen ordered a brain scan and "was able to point out to her the marked increase in activity in that area of her brain. It provided me with the evidence needed to convince her to go back on Prozac for a while longer." In this case, the scan images convinced the patient that the source of her poor mothering was biological and hence necessitated biologically based treatments. More important, the scans convinced her of the "reality" of her illness, ameliorating the stigma associated with problems that are thought to stem from character or personality. The change in interpretation produced by the scan changed what consuming medications meant to this patient. Moreover, when the patient consented to medical treatments, she was able to become "the good mother she wanted to be."[28] In this instance, the mother's refusal to take Prozac was a refusal of her own desire to be a good mother—seeing her scan and accepting medication allowed her to fulfill her own truest potential.

Medications are only one part of the treatment programs. Treatment is an ongoing process that includes a close attention to the moment-by-moment status of the brain. Images are used to evidence the biological effects of every activity and mood. Amen tells of one patient who was scanned twice: the first time, she was told to meditate on things that she was thankful for. In the second scan, she was told to think about all the things she hated about her life. The comparison of the scans allowed her to "see the difference that an attitude of gratitude can make in the brain." Amen warns, "Negative thought patterns change the brain in a negative way. Being grateful for the wonderful things in your life literally helps you have a brain to be grateful for" (MGBG, 151). Individuals must constantly tend to their thoughts, because literally "every thought" has an immediate, physical effect on the brain (MGBG, 152). Individuals must think about their thoughts, examining each one and assessing its positive or negative quality. This reflection should take the form of writing. Amen recommends, "Whenever you feel sad, mad, or nervous, write out what you are thinking. You will notice that many of those thoughts are irrational and hurtful" (MGBG, 153). Amen includes a specific worksheet, the "One-Page Miracle," in which readers are asked to write down their major goals using three headings (Relationships, Work/Finance, and Self). The paper is to be placed in a conspicuous location and consciously reflected on at least once a day. Like the more physical treatments of medication, these meditative and reflective practices are designed to heal the brain by allowing patients to access their own true selves and fulfill their own desires. Negative thoughts and ingratitude are brain-based, while contentment and gratitude are will-driven.

Amen's book is rife with different practices that are part of the brain treatment program, ranging from listening to specific types of music, filling out various worksheets, watching the movie Pollyanna regularly, and laughing often. These specifications of caring for the self are grounded in the visual authority of brain scans. The scans reveal the physical correlates of virtually all activities and are deployed as evidence of the ways in which these activities affect the brain. As the comparative brain scans demonstrating the value of an attitude of gratitude show, the brain can change in an instant in response to a mood or behavior. The functional brain scans support a treatment program that is continuous and ongoing, a lifelong project that is never complete. The access to the true self is never 100 percent, because the brain's function is always at risk of being thrown off balance by even the most inconsequential of life events. The true self, like brain health, must be constantly pursued and can be acquired in lesser and greater quantities, but its ultimate realization remains elusive.

Conclusion: Neuroscience and the Duty to Be Well

Despite optimistic promises of self-transformation, spiritual discovery, and worldly success, brain-based self-help sits a heavy burden on its readers' shoulders. Readers learn that in order to act on and achieve their deepest desires, they must first become their true selves. This is no easy task: the true self is buffeted on every side by internal and external forces that threaten its fragile balance. Because the true self depends on a brain that is not only structurally sound but also perfectly healthy in its function, it is a tenuous, elusive state rather than an identity one can hold with confidence. The self, in this sense, is internal because it somehow resides within individuals as the manifestation of their truest potential, but it is also something far, far away that must be achieved through ongoing, calculated projects. The various obligations to be well, articulated in various treatment programs, are grounded in a truth that is at once objective and technical, and derived from the individual's own internal, unique subjectivity. Individuals are governed through their own desires, their own interests in becoming healthy, fully whole subjects—they are, in Rose's terms, "obliged to be free." As Rose explains, these obligations to be free and well "are more profoundly subjectifying because they appear to emanate from our own individual desires to fulfill ourselves in our everyday lives, to craft our personalities, to discover who we really are."[29]

Amen's book and the phenomenon of brain-based self-help show how neuroscience discourses infiltrate life at the level of daily practices, working to shape the way individuals speak, think, and behave. The next chapter continues this discussion by looking at how brain images circulate in popular media about child care. Imaging of babies' brains has become an important component of child-care discourse over the past few years. These images work much like those in brain-based self-help, giving life a calculable form and encouraging individuals to act in the name of health. In the case of babies, individuals are encouraged to act not only for their own health, but also for the health of others, specifically children. One outcome is increasing support for social policies designed to cultivate and protect babies' brains. Self-fashioning through neuroscience is augmented by a certain type of social engineering that takes the brain as its focal point, framing babies' brains as resources and assessing social policy based on its ability to optimize these resources for different types of social and economic profit.

4

Babies, Blank Slates,
and Brain Building

The rhetorical consequences of the widespread dissemination of brain-based ways of thinking about life approached absurdity when, in January 1998, Zell Miller, the governor of Georgia at the time, entered the state legislature with a recording of Beethoven's Ninth Symphony. After playing a few minutes of "Ode to Joy," he asked the lawmakers, "Now don't you feel smarter already?"[1] Miller's dramatics were in support of his request that the legislature approve $105,000 to produce and distribute a classical music CD to parents of newborns throughout the state. Miller cited brain science research that indicated that listening to classical music enhanced mathematics and spatial reasoning abilities in babies. In the end, however, the allocation was unnecessary, as Sony agreed to provide the CDs free of charge; by that summer, parents of newborns in Georgia left the hospital with complimentary Mozart recordings.[2]

Miller was alternately praised and mocked for his "Mozart for Babies" initiative, which came the same year that Florida's government passed a bill requiring state-funded day-care centers to play classical music daily. Regardless of the political fallout from these particular incidents, however, the events were part of a much larger trend of attention to babies' brains and the intersections of neuroscience and public policy. The Florida initiative and Miller's proposal came about four years into the "Zero to Three" phenomenon, a loose affiliation of scientists, child advocates, celebrities, and politicians who insist that the first three years of an infant's life are crucial for the child's brain development and thus their future identity and behavior—as well as the fate of society as a whole. The attention to babies' brains is manifested in government-sponsored conferences, public awareness campaigns, popular media attention, and the explosion of products marketed to parents and educators, ranging from books and CDs to toys and clothing, that promise to aid in the task of "building" good baby brains. Although critics of the baby-brain movement's consumerist trajectory became increasingly vocal in the late 1990s and early 2000s, this criticism has largely served to intensify attention to babies' brains, shifting emphasis from cognitive-enhancing products to parenting and education practices designed to nurture the brain by attending to the "whole child."[3]

When Neuroscience and Politics Collide:
The Baby-Brain Movement

What is remarkable about the baby-brain movement is how such an enormous range of social practices is articulated as a neuroscientific issue. In the Carnegie Foundation's highly influential 1994 report *Starting Points*, the authors discuss unplanned pregnancies, the growing number of women in the workforce, divorce and the rise of single-parent families, poor child care, teenage delinquency, poverty, illiteracy, and the decline of "human capital" for the national workforce as issues directly related to neuroscience and informed by brain imaging evidence. Similarly, in his keynote address at the 1997 White House Conference on Early Child Development, the actor and director Rob Reiner stated that the zero-to-three theory is a way of "problem solving *at every level of society*."[4] He continued to drive home the point that focusing on the child's brain during this critical early period will have a positive impact on "children's success in school and later on in life, healthy relationships, but also an impact on reduction in crime, teen pregnancy, drug abuse, child abuse, welfare, homelessness, and a variety of other social ills."[5]

That conference was designed to "unite science with the everyday tasks of parenting and with public policy."[6] The Society for Neuroscience applauded the conference's attempt to "meld" neuroscience and public policy.[7] In 2001, Laura Bush hosted a similar conference, the White House Summit on Early Childhood Cognitive Development. Like the Clintons' 1997 conference, the Bush summit was a forum for scientists to disseminate research to public actors, including parents, educators, and policy makers. The summit emphasized the translation of cutting-edge scientific research on the infant brain into practical guidelines capable of informing the management of children at every level of society, from the family to the school to social welfare. In her introductory speech, Laura Bush stated that parents, educators, and other social actors share a "duty" to attend to and apply brain research, information "that everyone should know . . . not only mothers and fathers and caregivers, but also educators, health care professionals, policy makers, foundations, and businesses."[8]

These two White House conferences are compelling examples of the way in which neuroscience is increasingly viewed as a relevant language for speaking about a host of personal and public policy–related issues. The two conferences are strikingly similar in the types of language used to describe their purposes and frame their agendas. This congruence shows how the language of the brain functions as a nonpartisan discourse, free from any essential political or ideological dogma. The language of contemporary neuroscience participates in the apparent neutrality and objectivity often attributed to scientific discourse, providing a means of speaking the truth about family structure, education, and other social policies without ostensibly pushing an agenda.

The striking aspect of the spread of neuroscientific vocabularies is not simply that they provide an apparently neutral way to speak of political issues, but also that they inform social practices at multiple levels. More than just a body of research, neuroscience is a set of practical guidelines essential to living life in an optimal fashion. In the previous chapter, Amen's book was used as an example to illustrate how neuroscientific vocabularies frame the practice of living at the level of the individual. In the case of babies' brains, neuroscience is articulated as crucial to the practice of daily life for individuals, in their roles as parents and family members; for local institutions, including schools, educators, and social welfare agencies; and for larger collectives, including state and federal governments. Neuroscience informs the practice of life at every level—from the individual to the state— and across virtually every dimension of living, from education to family to hygiene to health care to entertainment. In their book *The First Three Years and Beyond: Brain Development and Social Policy*, Edward Zigler, Matia Finn-Stevenson, and Nancy Hall describe a host of issues that are informed by the neuroscience of young children, including early intervention and early education programs, nutrition during prenatal and early childhood years (including, for instance, WIC, the Women, Infants, and Children program), health care and health insurance, family cohesiveness, day care, welfare reform, and family and medical leave policies.[9]

Images of babies have long been a space to locate a society's hopes and fears, including both individual and collective dreams and nightmares. Babies signify innocence and possibility, and they are often featured as the centerpiece in discussions about the relative influence of nature and culture over human development and identity. The cover of a 1997 *Time* magazine special issue on babies and brain research encapsulates these notions of innocence and possibility. A baby stares from the cover, her intense look suggesting that she possesses some wisdom or insight despite her inexperience. The baby does not appear unhappy, nor does she appear to experience that happy abandon and naïve joy that mark many images of babies. Despite its seriousness, the image retains connotations of innocence, signified by the baby's pure white skin, open gaze, and expression of wonder. The baby's forehead is overlaid with another picture to suggest what possibilities might reside within the baby's mind. The picture is not of the baby's brain, which might be expected for a cover story about baby brain development. Instead, the cover depicts five figures, four very young children and one adult. One of the children beams out at the readers as she plays the violin. Everything about the young girl suggests that she is precocious, gifted, and socially adept. The other three children interact with one another and with the adult, intently engaged in activities with avid curiosity.

This *Time* cover image illustrates an important theme that courses throughout the baby-brain rhetorical formation. The overlaid picture of the children and the adult suggests that the possibilities depicted—of social interaction, development of musical talent, educational exceptionalism—somehow reside within the baby's mind. The baby's brain is more than just a biological organ—it is an object of public attention that can be cared for, manipulated, and interpreted through a variety of personal and public initiatives. Because everything from family interaction to education to social welfare can potentially influence the development of the baby's brain, it functions as a sort of rhetorical "space" that brings together a wide range of political, social, and economic issues and defines them in neuroscientific terminologies. Moreover, within this rhetorical space, causality goes both ways: social practices are powerful influences on babies' brains, just as babies' brains are framed as the foundation for the future, whether optimistically envisioned in terms of success, peace, and prosperity, or invoked in foreboding tones to warn of national decline, violence, and social decay. Brain imaging technologies play an important role in this assimilation of social life into the grammar of neuroscience because they visibly depict the biological correlates of a variety of social influences, ranging from medication to education to verbal interactions in the context of family. Taken with the familiar mantra that the brain is the ultimate cause of personality, attitudes, and behavior, the fusion of neuroscience and child care articulates babies' brains as a primary national resource that must be carefully cultivated in order to reap a wide range of benefits, including an educated citizenry, a thriving economy, and a peaceful society.

In part because of brain imaging technologies that visualize the biological dimensions of social factors and the social dimensions of the biological brain, the age-old "nature versus nurture" dichotomy is no longer a viable opposition. We increasingly understand culture as a biologically significant arena, defined in biological vocabularies, and on the other side of the same coin, we see the brain as something that is deeply dependent on sociocultural forces. In rhetorical terms, as a language of "speaking true" in the sense that Foucault uses these words, neuroscientific grammars are not reducible to nature; neuroscience is a useful vocabulary for expressing social

FIGURE 9 This is the cover of the February 3, 1997, *Time* special issue on the brain development of babies.

and cultural significance. In a way, what biological language refers to is not limited to nature, conventionally conceived—biological languages have referential capacity in the realms of family, politics, and culture, as seen in the previous chapter, where comments like "she's so cingulate" can circulate as colloquial descriptors of social behavior.

In the context of the management of families and state and local agencies, neuroscience plays a key role because it articulates babies' brains as objects of individual and social intervention, encouraging practices ranging from constant monitoring by family members to meticulously planned education initiatives. These practices, which range from the individual to the collective, exemplify a key characteristic of governmentality: a basic continuity in the exercise of power at different levels of society that draws together and intertwines ethics (self-care and self-management) with politics (regulation at the level of collectives, or the population). In short, the way individuals are encouraged to act in the context of their personal and family lives connects with broader social initiatives.

Despite the vocabulary of "interests," I want to be clear that I do not attribute a nefarious intent to any particular agent. This type of government is not the scheme of a central power, but more of a process of alignment that comes about through complex interactions that cannot be attributed to any single origin. Moreover, I do not want to suggest that individuals are simply tricked or duped when, guided by available discourses, they act out of their own perceived interests and desires for freedom, fulfillment, and empowerment. My purpose is simply to show that these interests and desires are never pure because they are constituted from available grammars that have social effects beyond the will or design of the individuals who adopt them.

In this chapter, I examine the baby-brain discourses through an analysis of *Time* and *Newsweek* cover stories from the late 1990s and early 2000s. The baby-brain discourses are built around a dialectic of permanence and malleability that is articulated through three major sets of tropes: metaphors of "wiring"; the concept of "windows of opportunity," or critical periods that define early child development; and, finally, the figure of "building" a baby's brain and, ultimately, a desirable society. These tropes are flexible linguistic resources that accommodate both determinism ("the brain is the cause of individual identity and the state of society") and plasticity ("individuals and societies shape the brain"), and this flexibility is key to the baby-brain discourses' rhetorical force. First, the baby-brain discourses draw on determinist languages, framing the brain as the cause of both a child's and society's future. This rhetorical vein draws from mechanistic metaphors to situate the brain as something that is permanently hardwired and, once built, largely resistant to change. Second, the languages of permanence are entwined with languages of malleability: babies' brains are uniquely plastic, vulnerable to all types of social intervention, and capable of extraordinary change. Change and permanence are organized temporally: there are unique "windows," or time periods when babies' brains are malleable, but after the windows shut, opportunities to build babies' brains cease and children are permanently destined to fulfill their biological destiny. Taken together, this assemblage of tropes mobilizes the fears and hopes associated with parenting to condition the proliferation of a host of interventions—welfare, education, and state governance of parenting—all designed to act at the level of babies' neurons to build a better society.

Time, *Newsweek*, and the Baby-Brain Rhetorical Formation

Child development has always been a controversial public issue, with the pendulum regularly swinging back and forth between nature and nurture. In the 1990s, however, a "paradigm shift" introduced the language of neuroscience to the discussion.[10] Brain imaging, according to Zigler, Finn-Stevenson, and Hall, "fundamentally changes the way we view our children and ourselves."[11] Brain imaging research is generally used to support the zero-to-three theory, or the belief that the first three years of a child's life are critical to shaping its brain, which will then come to determine its attitudes, behaviors, and experiences in later life. Scientific emphasis on the first three years of a child's life has a long history, although the nature of scientific advice for these critical years has varied widely throughout recent decades.[12] A critical centerpiece of the contemporary zero-to-three movement is the 1994 Carnegie report, *Starting Points*, that describes a "quiet crisis" in which children under the age of three "are in trouble, and their plight worsens everyday." The report cites brain imaging studies that suggest that the first three years are critical to child development. During this time period, brain development is "much more vulnerable to environmental influence," and this influence is "long lasting." The environment affects both the number of brain cells and the way they are "wired" into connections. The report concludes by calling for a bevy of social policy changes.[13]

The baby-brain discourses constitute a rhetorical formation that is highly intertextual, comprising the websites of nonprofit organizations, the public campaigns of celebrities, government pronouncements, and an array of media profiles. For instance, as mentioned above, the Carnegie report inspired Rob Reiner to develop a national public awareness campaign and create the I Am Your Child Foundation. Many celebrities, including Whoopi Goldberg, Tom Hanks, and Robin Williams, have been involved with the foundation's projects to gain publicity for issues related to children's neural development. The foundation is probably best known for its videos featuring celebrities addressing parents about basic child-care issues. Reiner also led a 1998 campaign to get Proposition 10 passed in California, legislation that increased the state tax on tobacco products to fund early childhood development programs. Other states have been involved in similar initiatives to support early childhood development in the wake of the baby-brain enthusiasm.

The most important distribution points in the baby-brain rhetorical formation are parenting advice books, advertisements, public policy events, and news stories. Wendy Cole describes the consequences of the "media blitz" of a cause embraced by celebrities and schoolteachers alike: "Every new mom I knew was rushing out to buy the latest in high-contrast black-and-white toys purported to stimulate neurological development."[14] Stephen Hall of the *New York Times Magazine* writes that the media attention to baby brains has resulted in "a neurotic national pastime: raising a scientifically correct child."[15] In their analysis of media coverage on the topic, Zigler, Finn-Stevenson, and Hall conclude that brain-based child development stories "do seem to have caused a shift in how parents perceive both the nature of early development and their role in fostering it."[16] Surveys in conjunction with the 1997 White House conference suggest that 92 percent of parents believe that experiences before age three will influence children's success in school; 85 percent believe that without appropriate stimulation, children's brains will not develop properly; and 60 percent responded that they were extremely or very interested in learning more about brain research.[17] The media coverage has found a "large and receptive audience," which of course influences the selection choices for news and feature coverage.[18]

I focus on two cover stories from *Newsweek*, and two special issues, one from *Time* and one from *Newsweek*. I have selected these texts on the premise that mainstream news magazines reach wide audiences and are consistent with broader rhetorical trends, especially because magazine stories frequently include tidbits from a wide range of discourses, including scientific studies, popular advice books, and other news stories. Moreover, statistical and anecdotal evidence suggests that these particular texts have been especially influential. The first story is Sharon Begley's cover story for the February 19, 1996, issue of *Newsweek*, titled "Your Child's Brain: How Kids Are Wired for Music, Math, and Emotions." Begley's article "brought the new brain science and its potential implications for early childhood to mainstream America and the world."[19] The public reaction to the article was "overwhelming," as *Newsweek* received more reprint requests for the article than it ever had for other any article, and the issue was one of the year's best sellers.[20] The success of this story eventually led *Newsweek* to publish a special issue in 1997, titled "Your Child: From Birth to Three," for which the magazine worked with Reiner and the I Am Your Child Foundation. It was a "massive success," selling around one million copies and going through several printings.[21] In addition to the *Newsweek* story, I selected two additional stories from popular magazines that illuminate the discursive trends of the baby-brain movement. The other *Newsweek* story selected is the more recent "Your Baby's Brain" from 2005. In addition, I examine a 1997 issue of *Time* magazine, titled "How a Child's Brain Develops and What It Means for Childcare and Welfare Reform."

Child development has always been a hot topic, so the prominence of the baby-brain research in public discourse is not difficult to understand. The baby-brain discourses are popular in part because they foster both a sense of guilt and a sense of control that fuel increased consumption of parenting advice discourses. On the one hand, these discourses generate guilt and anxiety in parents by fostering fear that they not doing enough, or are doing too much of the wrong thing, thereby compromising their child's future. Simultaneously, the discourses generate a sense of hope by telling parents that their actions have enormous influence for their child's future development, and that they can bring about desired outcomes with the right information, the right products, and the right practices. As suggested by the *Time* cover image featuring older children overlaid on a baby's head, baby-brain rhetorics tap into parents' hopes for their children's future success in terms of education, finances, and social status. As I will explain, these hopes for individual children's futures are often expressly linked to hopes for social redemption, including national economic competitiveness, global status, and domestic peace. In the next section, I examine how wiring and window tropes work together to structure a dialectic of permanence and malleability, contributing to this ambivalent construction of guilt and hope.

The Baby's Brain: Wired and Windowed

Although various aspects of the zero-to-three theory are contested in the baby-brain discourses, what is taken for granted by almost all participants in the debate is that neuroscience research should inform public policy and guide caretakers, including parents and educators, in their daily interactions with children. For instance, Matthew Melmed, executive director of the nonprofit organization Zero to Three, writes to *Newsweek*, "What parents need is guidance on how to apply all this new knowledge to support their child's development through everyday interactions."[22] In a *Time* cover story challenging the zero-to-three theory's tendency to promote the frenzied consumption of cognitive-enhancement products, Jeffrey Kluger and Alice Park suggest that science actually indicates that parents need to "relax" and pay more attention to emotional attunement and positive social interactions with their children, and less attention to the latest products marketed by zero-to-three-influenced "hucksters."[23] Even critics of various manifestations of zero-to-three largely accept the assumptions and the terminologies of the discourse. Two clusters of tropes constitute this rhetorical configuration: technological metaphors, which describe the infant's brain as something that is wired, and the ubiquitous discussion of windows and critical periods.

Scientific language is, in one sense, a series of metaphors that draws on culturally relevant discourses to understand natural phenomena.[24] Metaphors in public discourse are deeply influenced by context, and both the selection and the meaning of metaphors are dependent on contingent cultural and historical features. Steven Montgomery writes that the force of metaphors "lies in their ability to create images or even whole image systems."[25] Throughout the baby-brain discourses, the imagistic verbal metaphors operate amid a variety of visual images. These visual elements are important contextual features of the baby-brain discourses and, in some cases, play a more important role than the verbal features of the text. As I describe the three sets of verbal metaphors that define this discourse, I will refer to the three major categories of visual images that accompany this copy: images of babies, functional images of babies' brains, and, finally, charts and diagrams. These visual elements rarely appear alone, and in most cases all three visual features appear in some type of combination or juxtaposition.

Technological Metaphors: Wiring the Baby's Brain

The language of "wiring" is a powerful metaphor for framing babies' brains, in part because it is able to activate meanings associated with both determinist claims and rhetorics of plasticity. When the brain is "hardwired," it is often talked about in mechanistic metaphors, an assemblage of tropes that has been a common feature in scientific discussions ever since Descartes compared the human body to a clock. The invention of the computer added a new set of terms to the scientific repertoire, providing scientists of the mind with a handy vocabulary for describing seemingly intangible processes by way of the concrete. In general, mechanistic metaphors tend to see the brain as a machine or system, with its functional processes set by biology. Even though mechanistic brains can be made to produce different outcomes when different inputs or "software" are introduced, the machinelike brain tends to be framed as something relatively set or hardwired. In addition to the mechanistic meanings, wiring metaphors are also able to accommodate paradigms that view the brain in terms of connectivity and dynamic plasticity. Donna Haraway argues that the major movement defining the "paradigm shift" in the life sciences in the past century is "an effort to deal with systems and their transformations in time," utilizing mobile and dynamic metaphors to describe the function of living systems.[26] The wiring metaphors can activate a sense of movement and process to various degrees, depending on the contexts of usage.

The more mechanistic meanings of wiring metaphors are activated when they are used as an extension of, or in combination with, computer metaphors. At birth, certain parts of the baby's brain are "hardwired," or already determined by nature. Other parts exist as an indeterminate mass of neurons that have not been arranged into a functional structure. Cultural influence creates and reinforces connections (synapses) among these neurons, effectively "wiring" the brain into a determinate structure of organized circuits. The wiring metaphors enable ambivalent meanings about the respective roles of natural and cultural influence. Although parts of the brain are hardwired, cultural agency seemingly has wide berth in affecting the rest of the brain. But cultural forces must tread carefully because once the brain is wired, it is wired for good, becoming a permanent biological structure that controls the rest of the child's life.

In her 1996 *Newsweek* cover story "Your Child's Brain," Begley describes the wiring of the baby's brain by way of a computer analogy. Babies come into the world with a "jumble of neurons," some of which are "hardwired" into circuits that control breathing, heartbeat, and other basic motor functions. Most neurons, however, are not:

> Trillions upon trillions more are like the Pentium chips in a computer before the factory preloads the software. They are pure and of almost infinite potential, unprogrammed circuits that might one day compose rap songs and do calculus, erupt in fury and melt in ecstasy. If the neurons are used, they become integrated into the circuitry of the brain by connecting to other neurons; if they are not used, they may die. It is the experiences of childhood, determining which neurons are used, that wire the circuits of the brain as surely as a programmer at a keyboard reconfigures the circuits in a computer. Which keys are typed—which experiences a child has—determines whether the child grows up to be intelligent or dull, fearful or self-assured, articulate or tongue-tied. Early experiences are so powerful, says pediatric neurobiologist Harry Chugani of Wayne State University, that "they can completely change the way a person turns out."[27]

The description of neurons as Pentium chips that are "pure and of almost infinite potential" taps into long-standing beliefs about the purity and innocence of childhood. Despite scientific aversion to the "blank slate" doctrine, an element of it persists in the technological metaphors of brain wiring.[28] Descriptions of the infant's brain construct its purity and "infinite potential" by emphasizing the sheer quantity of neural cells. Madeleine Nash writes that at birth, "the baby's brain contains 100 billion neurons, roughly as many nerve cells as there are stars in the Milky Way."[29] This emphasis on quantity depicts the infant's brain as simultaneously unlimited and biological. "Nature" theories are often thought of as more limiting than nurture theories, because they suggest that identity is constrained by biology. The description of "trillions and trillions" of neurons makes nature something that is itself infinite, a type of "unprogrammed" tabula rasa of neurons. The blank slate is replaced with the unprogrammed computer that has not yet been "wired" into the circuits that will then determine its functions.

The computer metaphor enables a complex interaction between natural and cultural agencies. The unprogrammed brain exists as passive matter, awaiting the imprint of form from active cultural agencies. Once these circuits are formatted, however, they become the forces that determine the computer's functions. The stark contrast between the possible programming outcomes is illustrated by the description of extremes: the child can experience "fury" or "ecstasy"; its neurons can become "integrated" and "connected," or they can "die"; the child can become "intelligent or dull," "fearful or self-assured," or "articulate or tongue-tied." These oppositions, and the added emphasis that early influences can "completely change" a child's future, retain elements of a biological determinism, or "nature" perspective. Once the biological circuits are "determined," the child's fate is preordained. In this sense, biology truly is destiny.

This construction of opposites not only emphasizes the permanence and power of the child's biology, it also attests to the power of cultural agency. Biology might determine whether a child is a failure or a success as an adult, but this biology is itself determined by cultural influence. Experience is analogous to a computer programmer, the agent who establishes and orders the connections that will later determine the functions and behaviors of the machine. The description of a programmer who systematically and rationally sets out this circuitry by typing the appropriate keys suggests that this cultural agency functions according to an accessible cause-and-effect logic. John Bruer suggests that these metaphors are why the baby-brain discourses appeal not only to women but also to men: men are attracted to the mechanistic construction of child development as something logical and systematic.[30]

Two common alternatives to computer wiring are the telephone wiring and the electrical wiring metaphors. When the language of wiring is deployed in these contexts, it tends to activate meanings more closely associated with theories of plasticity and malleability. The brain is not so much a machine as a dynamic system of connections that are constantly being created, renewed, and destroyed. The developing nervous system "has strung the equivalent of telephone trunk lines between the right neighborhoods and the right cities. Now it has to sort out which wires belong to which house, a problem that cannot be solved by genes alone."[31] Connections are formed through experience, as connections that are used are reinforced and those that are neglected die off and wither away. The process of forming connections is like "teenagers with telephones, cells in one neighborhood of the brain are calling friends in another, and these cells are calling their friends, and they keep calling one another over and over again."[32] For calls that are made frequently, the paths are preserved as the connections become myelinated, covered with a white, fatty substance "that coats nerve cells like the plastic insulation on telephone wires" and keeps signals on track, preventing "cross-talk."[33]

The telephone metaphors situate the baby's brain as a dispersed set of "neighborhoods," or different areas with different functions that must coordinate in order to function properly. The brain, in this sense, is an array of distributed systems that operate through interaction and coordination. Just like telephones appear to ameliorate distance and bring dispersed individuals into what seems like immediate contact, brain signaling similarly eliminates the space of the brain through signaling processes. The brain is not a "thing," but a set of combinations or interactive dynamic functions. The functional brain images reinforce this perspective. These images are colored representations of brain activity, and they are very different from images of the "wet brain" found in older scientific texts. While the wet brain looks like a single organ, the functional brain images show a dynamic grid of activity with different shadings that suggest an ability to change given a simple alteration in stimulus. The functional brain images look more "real" or authentic than images of the wet brain, which have little ability to suggest the almost magical processes associated with the organ.

Communication between neighborhoods in the brain depends on proper signaling systems, and these are formed in practice as linkages biologically preserved because of their frequent usage. This communication metaphor is tied to the biological theory that neurons communicate across synaptic gaps through electrical and chemical means. The idea of a message being encoded and transmitted across long distances through telephone wires is close to the idea of electrical signals transmitted throughout the brain. The electrical connections that are formed are often described through auditory language. LynNell Hancock and Pat Wingert begin their *Newsweek* article on brain wiring by telling the reader to "listen to the snap, crackle, pop of baby neurons."[34] Barbara Kantrowitz similarly writes, "Every lullaby, every giggle and peek-a-boo, triggers a crackling along his neural pathways, laying the groundwork for what could someday be a love of art or a talent for soccer or a gift for making and keeping friends."[35]

These wiring metaphors suggest that the linkages or connections are formed through social experience immediately, in the instant it takes to hear another's voice on the line. Social agency constantly produces biological effects, and there is no temporal gap between the stimulus and the wiring response. Brain imaging is frequently cited as evidence of the coincidence of stimulus and response. Begley writes:

> You cannot see what is going on inside your newborn's brain. You cannot see the electrical activity as her eyes lock onto yours and, almost instantaneously, a neuron in her retina makes a connection to one in her brain's visual cortex that will last all her life. The image of your face has become an enduring memory in her mind. And you cannot see the explosive release of a neurotransmitter—brain chemical— as a neuron from your baby's ear, carrying the electrically encoded sound of "ma," connects to a neuron in her auditory cortex. "Ma" has now commandeered a cluster of cells in the infant's brain that will, as long as the child lives, respond to no other sound. You cannot see any of this. But Dr. Harry Chugani can come close. With positron-emission tomography, Chugani . . . watches the regions of a baby's brain turn on, one after another, like city neighborhoods having their electricity restored after a blackout.[36]

Chugani can "measure" brain activity and "observe" the cortex "burn with activity" and "light up" as experiences "determine the actual wiring" of the infant's brain.[37]

These visual vocabularies are accompanied by two images of PET scans, clearly marked as different by their contrasting colors (one is primarily blue while the other is primarily red) and by the labels "Healthy Brain" and "Abused Brain."[38] The color scale indicates to the reader that the healthy brain, the red brain that is "glowing" or "lit up," shows high activity while the blue brain exhibits low activity, evidence of "extreme deprivation." These visual vocabularies and the accompanying images suggest that brain imaging provides science with moment-by-moment access to the brain, and that the effects of a particular stimulus can be observed instantly. The wiring process is instantaneous, but the effects are long lasting. The actual wiring is "determined," and the connections described above will last the entirety of the baby's life. Sight, in this case, provides not only constant access that shows that social stimuli have immediate effects, but also is taken as something that accesses "nature." The brain images are thus ambivalent: they are used as evidence of the susceptibility of the brain to cultural influence, and they are also used as a representation of a natural referent, invoking the qualities of permanence and immutability associated with biological determinism.

Not only are the effects of social stimuli instantaneous and permanent, but they can be brought about through the most casual interactions. The wiring process happens literally in the blink of an eye: speaking a particular word, touching in a particular way, even once, can have lasting effects. If the brain is a blank slate, it does not take a heavy hand to mark it permanently. As Kantrowitz's story states, every lullaby, every interaction, solders connections affecting the structure of the brain for life. The consequence is that it is possible to build a baby's brain inadvertently. Debra Rosenberg writes in *Newsweek*, "Parents may be unintentionally sending signals from the start, or deliberately shaping the most crucial messages."[39] The metaphor of the computer, where the brain is affected by every keystroke, is supplemented by the telephone and electric circuitry metaphors, which suggest that connections can be made immediately and unintentionally. If the wrong stimulus is presented at the wrong time, it is akin to dialing a wrong number: the circuit is still activated, even if it was an accident. This undercuts the agency implied in the mechanistic metaphors of building a baby's brain. Social influence is substantial, but it is not governed by intentionality. Chance and accident are just as likely contributors to the wiring process. A single incident, such as a mother who screams at her child, or a father who arrives home drunk and beats his child, can create pathways with lasting effects (e.g., "the mere memory of Dad may induce fear").[40]

The wiring metaphors interact with the images of babies in contact with imaging technologies. The 2005 *Newsweek* cover, for instance, features a white, blue-eyed baby gazing in marvel at the wires that dangle from the geodesic sensor net attached to his head. The net consists of a number of small "suction cups" that are designed to measure the electrical activity of the brain. The image combines a socially ideal baby (white, blue-eyed, curious, healthy) with technology that has a science-fiction connotation: a net of wires connected to the head, suggesting a certain spillover between baby and machine. This image combines the innocence of childhood, with its representation of a perfect baby, and the promises of biotechnology. The baby appears contented, even fascinated, by the geodesic sensor net, suggesting that the imaging technologies are truly noninvasive. The sensors attach to the baby's head, but they do not penetrate his skin. The baby remains intact, despite the revelations that are enabled by the net. In this image, the baby is in part a synecdochical representation of humanity, and more specifically, the capacities of the human mind in general.

Taken together, the wiring metaphors suggest a complex interaction between nature and nurture. At birth, the infant is partially determined (hardwired) by nature. The rest of the infant is passive matter, a jumble of potential connections, ready to be wired into a circuit. From there, social agency takes over, establishing connections through interactions. But social agency is not unbound. Nurture must follow the hidden rules of nature, providing the right influence at the right time to establish connections that will produce a good child. To fall short of this hidden guide, to establish connections through accident, caprice, or error, is fatal: literally, to the neurons that

FIGURE 10 This photo appeared on the cover of *Newsweek*, August 15, 2005, accompanied by the headline "Your Baby's Brain: From Jealousy to Joy: How Science Is Unlocking the Inner Lives of Infants."

Timothy Archibald/Timothy Archibald Photography

will die, and metaphorically, to the opportunity to build a successful (or normal) child. What counts as success or normality from a biological standpoint is ultimately judged only by its social manifestation: if the child turns out to be dull and inarticulate, it is inferred that the child's brain was not properly wired (not exposed to the right influences) during its first three years.

Timing Is Everything: Critical Periods and Windows of Opportunity

The combination of natural and social agencies that "wire" the infant brain is structured by a strict temporal logic described in the rhetoric of "windows of opportunity," "milestones," and "critical periods." These are time windows when the brain is receptive to acquiring certain types of information; in fact, it needs specific stimuli during these periods in order to develop properly. The baby's brain is wired during the first three years of its life, but these three years are subdivided into a number of different time periods during which appropriate social influences are critical. These windows are described throughout the baby-brain discourses, both in textual accounts and in numerous charts, checklists, and timelines for tracking proper infant development. The metaphor has three important connotations. First, the critical periods are windows of access, providing an opening during which parents can shape and control the brain through appropriate stimulation. Second, they are windows of opportunity, in the sense that they are "once in a lifetime" chances to get things right. In this sense, they are windows that can be shut permanently, and therefore potentially signal missed opportunities. Finally, the language of windows is a visual metaphor that works in tandem with the visual images of the brain produced by technologies that are themselves frequently described as windows.

The windows metaphor situates these time periods as crucial because, like actual windows, they can be closed. Once the windows are closed, the wiring is in place and the future course is set. As Begley describes it, "Once wired, there are limits to the brain's ability to create itself. Time limits. Called 'critical periods,' they are windows of opportunity that nature flings open, starting before birth, and then slams shut one by one, with every additional candle on the child's birthday cake."[41] The stark description of windows "slamming shut" suggests that the time period is strictly defined, and that a missed opportunity is missed forever. The consequences of a missed window can be devastating. Madeleine Nash writes of brain imaging studies that show that the "emotional tone" of exchanges between mothers and their children determines the child's later emotional intelligence. The studies found that "mothers who were disengaged, irritable, or impatient had babies with sad brains." Timing is everything: for a short period, the baby's brain is "forgiving" and emotional damage can be repaired. "If a mother snaps out of her depression before her child is a year old," brain activity picks up. If the mother remains depressed, then the window is lost. The baby's first years are marked by "critical or sensitive periods," "when the brain demands certain type of input in order to create or stabilize certain long-lasting structures."[42]

These biologically defined timetables have two major discursive consequences. First, they construct the baby's brain as extremely vulnerable. The "open windows" make the baby's brain a sponge of sorts, which can be permanently affected by any exposure. Any stimuli, accidental or intentional, can potentially have adverse consequences for the baby's life. Second, the window metaphors define infant brain development as something that is determined by biology but requires precisely timed social input to stay on course. If the right stimulus is not given at the right time, the baby's brain is not wired according to nature's plan. Nash's article, for instance, describes nature and nurture as a "dance," in which nature is the dominant partner but nurture plays a vital supporting role. An absence of the necessary social "triggers," or social interference of the wrong type, can hijack the "clockwork precision of the neural assembly line."[43] The consequences are permanent. According to Dr. Bruce Perry of Baylor College of Medicine, children exposed to trauma and unpredictable stress, such as a mother's boyfriend who lashes out in anger, will suffer permanent consequences: "Some percentage of capacity is lost. A piece of the child is lost forever."[44]

Other descriptions similarly use quantitative figures to express the consequences of improper stimulation. Both Nash and Begley report that a child's brain suffers when deprived of a stimulating environment and is 20 to 30 percent smaller than brains of normal children.[45] The brains of deprived children also have "fewer synapses."[46] Through these articulations, not only brain health but also social interactivity is rendered quantitative and is assigned a numerical value. The visual brain images contribute to this quantitative sense, depicting contrasting images of the healthy brain and the sick or "sad" brain that does not receive adequate simulation. The different shadings and areas of activity depicted suggest a proportional relation between three factors: the amount of good social influence, the size of the brain, and the future success of the child.

The baby's brain, then, is "actively vulnerable," or vulnerable not only to trauma but also to a lack of proper stimuli. Nature and nurture are locked in a dance of "tightly choreographed steps," and if nurture fails to follow nature's lead, tragedy ensues. The baby-brain stories are peppered with diagrams, checklists, and charts that visualize this "dance." In general, these graphics communicate three major things. First, they illustrate what activities are expected, or "normal," at what age. The 1997 *Newsweek* special issue includes several of these charts, including "Growing Up, Step by Step," a series of graphics that "track an average child's development from zero to three."[47] For instance, readers learn that at eleven months, the baby "likes to turn pages, often not one by one. Fascinated by hinges and may swing door back and forth." Between thirty and thirty-six months, the child "rotates jigsaw pieces and completes a simple puzzle," and "tries out new types of movement like galloping and trotting."[48] These types of step-by-step guides function as a checklist, allowing parents to track their child's development and compare it to the "average" child, watching for symptoms of abnormal development.

Second, these graphics communicate what interventions are appropriate at what times. The 2005 *Newsweek* cover story includes a diagram titled "Milestones" for parents to "help track [their] baby's progress in relating to others, along with activities to help him meet these targets." The chart has three rows, dedicated to "emotions," "social skills," and "helpful games," a list of interventions that will help to assure proper emotional development.

For a baby of five to six months, for example, the parent should "use words and funny facial expressions to get [the] baby to break into a big smile."[49] Throughout these articles, the diagrams vary slightly in terms of format, focus, and specificity. But they are all designed to provide a foundation for tracking infant development and to prescribe appropriate types of influence for each critical period.

The graphics' final communicative function is to make readers literate in the language of neuroscience, enabling parents to use biological vocabularies to understand and interpret their child's development. A regular feature of the diagrams is visual representations. The articles often include diagrams of the baby's biological brain, allowing readers to connect the critical period, the appropriate functions for that period, and the exact area of the brain that is responsible for those functions. The 2005 *Newsweek* article features a photograph of a smiling baby, with a diagram of brain biology overlaid on the baby's head. The relevant areas are highlighted and an arrow connects the area to textual description. The diagram identifies the "dorsolateral prefrontal cortex," for instance, and tells readers, "This area may help babies remember people and things that aren't there. Once it's developed, babies can suffer separation anxiety over absent parents. Sees big gains after eight months." In some cases, these descriptions reference imaging studies. For instance, the discussion of the "left temporal lobe" states, "Brain scans: From as early as nine months, differences in temperament are reflected in brain activity. Shy babies show heightened activity in their left frontal lobes."[50]

The metaphors of wiring and windows function to divide natural and cultural agencies through temporal boundaries. The first three years of life are marked into "critical periods," temporal boundaries that illustrate normative natural development and provide a framework for precisely timed social interventions. The metaphor of "wiring" suggests that once neural connections are established, or linked by way of social interaction, they are then determined as a stable, permanent structure. The control of cultural influence is total in its consequences but not in its intentionality. It can make the difference between a successful child and a child with severe physical, emotional, and cognitive problems, but culture cannot rationally ordain desired outcomes. The first three years are governed by precise rules of timing that are only partially accessible to cultural agents. Virtually every experience, intended or accidental, can permanently wire the brain. Brain imaging research can access these rules of timing, but a child's brain cannot be permanently imaged throughout its first three years. To substitute for this scientific vision, depictions of babies in electrode caps or in fMRI machines remind readers that every baby has a wealth of data inside its body. Because it is not practical to image every baby's brain all the time, parents can partially substitute for vision by adopting a precise neuroscience vocabulary and attending to the behavioral details that are constructed as potential symptoms or evidence of abnormal development. In the next section, I will continue this analysis, describing how the images and metaphors of wiring and windows work together with languages of "building" babies' brains to authorize wide-ranging social policies and government interventions.

Technologies of Intervention: "Building" a Good Baby Brain

The 1994 Carnegie report, in many ways the starting point of the American obsession with babies' brains, describes the problem in stark terms: "Our nation's children under the age of three and their families are in trouble, and their plight worsens every day." The good news is that "given sufficient focus and sufficient political will, America can begin to find its way toward solutions. Our nation can formulate and implement social policy that responds, over time, to the most urgent needs of our youngest children." Children under the age of three "need our compassion and our help, and we, as a nation, have an incalculable stake in their well-being."[51] As the above excerpts illustrate, the problems of early child development are defined as biological or neuroscientific problems, but they are also social problems that affect the entire nation. Responses to the threat of abnormal biological development must come not at the level of individual biological remedy, but at the level of national policy making and social action. As Nikolas Rose explains, "In the name of public citizenship *and* private welfare, the family has been configured as a matrix for organizing domestic, conjugal, and child-rearing arrangements and instrumentalizing wage labour and consumption. In the name of social *and* personal wellbeing, a complex apparatus of health and therapeutics has been assembled, concerned with the management of the individual and social body as a vital national resource, and the management of 'problems of living,' made up of techniques of advice and guidance, medics, clinics, guides, and counselors."[52] Rose's conclusion is that what counts as "politics" is not self-evident in this age of governmentality, but must itself be an object of analysis. In the baby-brain discourses, the most "private" modes of caring for the self—for instance, the management of one's emotions in the home—become articulated with, or discursively linked to, public policy, including federally funded programs that mandate family leave, fund preschool education, and reform welfare.

A few examples of the public policy initiatives that emerge from this discursive configuration better illustrate these features of governmentality. The 1997 *Time* special issue on child brain development includes a graph that maps what different states are doing to address the "quiet crisis." North

Carolina, for instance, has "Smart Start," a program in which "parents, teachers, doctors and nurses, child-care providers, ministers, and businesspeople form partnerships at the county level that set goals for the education and health care of children under six." The article quotes North Carolina Governor James Hunt, who relies on the "hard science" of brain imaging research to bolster his policy initiatives. He says of the sensitivity of babies' brains, "Now that we can measure it and prove it, and if it can be made widely known so people understand this, then they'll understand why their schools aren't going to work for them, their technical training isn't going to work, other things we do later on aren't going to work fully unless we do this part right and do this at the appropriate time."[53]

Other states similarly use "hard science" to support initiatives targeted at young children. Oregon has "Healthy Start," a program that funds home visits to check up on child development. The article describes a twenty-two-year-old mother who was told by a home visitor that she should start reading to her child immediately, not to wait until the child was two or three years old. Vermont has Success by Six, a program that visits a home within the first two weeks of the baby's birth. The article quotes former Governor Howard Dean as saying, "That gets us in the door at age zero instead of age five, so we can assess what families need."[54] In a letter to *Newsweek* in response to the 2005 cover story on babies' brains, three professionals who work for Babies Can't Wait, a federal- and state-funded program that provides a "free, full-developmental assessment for any baby, up to age three," describe their own initiative. They write, "If the baby is found significantly delayed in any area of development, he or she is plugged into a system that will provide a home-based early-intervention program to address specific goals for that child. We are identifying babies as early as a few weeks old who appear to have signs of emotional, sensory, or behavioral problems."[55]

On the one hand, these interventions sponsored by federal, state, and local governments appear to represent an intrusion of state power into the private sphere. What is key from the perspective of an analytics of government, however, is the ways in which the functions of power at the level of the state—early intervention, diagnosis, monitoring—merge with the functions of individuals, families, and other private actors such as physicians and caregivers. As North Carolina's Smart Start program shows, the state is less the "owner" of power who intrudes in the lives of private citizens than a space of distribution or a mode of coordinating the free-floating mechanisms of regulation that are taken up by diverse social agents. With governmentality, the state's role is "one that gathers together disparate technologies of governing inhabiting many sites."[56] The relation between the state and the citizen is not one of domination or opposition, but one of enabling and assisting. As in Dean's quote, the government will assess what families *need*, responding to their desires for assistance. The policies that emerge from the baby discourses are "progressive," in the sense that they promote family leave and child care, typically support more generous and less restrictive welfare policies, and fund educational initiatives. In short, they give money to "help" babies, children, and families. It is this "progressive" nature of the interventions that poses serious rhetorical challenges to opponents of the "myth" of the first three years.

Conclusion

Brain imaging supports a powerful rhetoric that impels individuals to act on both their own lives and the lives of their children with a sense of urgency. Because of the brain's vulnerability to social influences and its obscure developmental timing sequence, parents and other social actors are encouraged to see babies' brains as resources that must be attended to in very particular ways in order to ensure optimal future rewards. Babies' brains are constructed as both the digital complement to the blank slate—open-ended potential that must be "wired" into fixed form—and as complex systems that demand precisely timed and predetermined interventions. Social actors, particularly parents and educators, are required to constantly interrogate how their moods and behaviors are permanently influencing babies' brains. This is not only an interrogation of the moods and behaviors that are enacted, but also of the activities that are not produced. The failure to engage in an activity at the proper time can be just as detrimental as engaging in the wrong activity. The oppositions between nature and nurture become entangled in this model that posits both malleability and permanence, both a tabula rasa ideology and the trappings of biological determinism.

Like the self-help discourses analyzed in the previous chapter, the baby-brain rhetorics are spoken in heady languages privileging social agency. Through scientifically validated parenting practices, education policy, and social welfare programs, society itself can be created anew as babies are built into productive, peaceful, and prosperous citizens. But this utopian promise comes at a steep cost. The flip side of unrestrained social agency is relentless responsibility. Bad parenting (as well as inadequate education and welfare programs) causes bad baby brains, and bad, or unhealthy, baby brains cause bad societies. Further, while babies' brains might be raw materials that can be molded or built through social agencies, the laws governing this building process are opaque. An intricate logic of timing governs the brain-building process, and too little or too much at the wrong time can have lasting consequences, causing permanent damage that will manifest as serious social problems, ranging from intellectual mediocrity to poverty and violence. Popular neuroscience's ability to frame the brain as simultaneously the ultimate cause and the most important effect of individual and social behaviors makes it a powerful rhetoric for articulating life as an ongoing, calculable project demanding constant scrutiny and ceaseless individual and social interventions. In the next chapter, I continue to explore the diverse consequences of popular neuroscience's treatment of life and health by examining a museum exhibit that blends marketing, education, and entertainment.

5

Pills, Power, and the
Neuroscience of Everyday Life

One of the most controversial aspects of brain culture is the prevalence of psychiatric medications, and especially the escalating consumption of these medications by children. In recent years, a number of advocacy organizations have emerged to vocally protest the "drugging" of U.S. children and adults. For example, CHAADA (Children and Adults Against Drugging America) was founded in December 2005 to expose and condemn the "conspiracies of corporate control over the mental health system." The membership organization encourages parents to refuse to "drug and poison" their children "in the name of treating diseases, like ADHD, that were merely invented for profit."[1] A similar group, Fight for Kids, was formed as an offshoot of the Citizens Commission on Human Rights, an anti-psychiatry group created in 1969 through the collaborative efforts of the Church of Scientology and Dr. Thomas Szasz, one of the most outspoken critics of psychiatric diagnoses and medications. Fight for Kids works to expose the "huge sales, public relations, and propaganda push" that function to "deceive and overwhelm the cautionary instincts of millions of parents and teachers."[2] These advocacy groups are just one small part of a multifaceted critique of the United States' affinity for psychotropic remedies that circulates throughout popular media, as further evidenced by the flurry of books published in the early 2000s—including those with titles like *Our Daily Meds: How the Pharmaceutical Companies Transformed Themselves into Slick Marketing Machines and Hooked the Nation on Prescription Drugs*; *The Truth about Drug Companies: How They Deceive Us and What to Do about It*; and *America Fooled: The Truth about Antidepressants, Antipsychotics, and How We've Been Deceived*—that target pharmaceutical companies for their role in perpetuating the drugging of the American public.[3]

These discourses share the belief that the popularity of psychiatric drugs is the result of manipulation and deception perpetrated by powerful economic interests. Pharmaceutical companies are primary targets of criticism, but psychiatrists and federal agencies, including the Food and Drug Administration and National Institute of Mental Health, are also identified as guilty parties. While I share with these critics an attitude of suspicion and skepticism toward pharmaceutical companies, my take is somewhat different as I attribute this popularity to far-reaching cultural and rhetorical processes, and not solely to greedy pharmaceutical corporations. As Gary Greenburg explains, "The captains of the pharmaceutical industry are merely doing what they get paid the big bucks to do—to sail their corporate ships expertly on the winds and currents of the times."[4] The "winds and currents of the times" include the widespread belief that unhappiness, as well as other undesirable moods and behaviors, are illnesses that can be treated through biomedical interventions. While pharmaceutical companies reinforce this belief through their marketing strategies, it is at the same time a prerequisite for the success of their message and their products—in short, pharmaceutical corporations both contribute to and react to the common cultural conviction that the brain and its chemical processes are the fundamental source of personal and social success (or failure, as the case may be). At the root of this conviction is the idea that happiness and health are fundamentally linked, and possibly even equivalent. Americans' concern for happiness is not new—as Greenburg notes, the question "Am I happy enough?" has been "a staple of American self-reflection since Thomas Jefferson declared ours the first country on earth dedicated to the pursuit of happiness." What is relatively new, however, is the question "Am I not happy enough because I am sick?"[5] In other words, the success of pharmaceuticals is part of a more general process whereby happiness is defined in healthist terms and pursued through biomedical interventions. Specifically, in the context of psychiatry, the rise of chemical imbalance theories of depression and other mental problems have been particularly powerful in shaping such healthist understandings of experience and facilitating the extraordinary success of psychotropic drugs.

In this chapter, I argue that the success of psychotropic drugs is best understood as an effect of the saturation of public discourse with biological and neurological ways of thinking. Following Emily Martin, I use the notion of saturation to "capture the sense in which awareness of and regard for the body's health," as defined by its brain function, "have come to be so general in the society that one cannot avoid it, wherever one turns."[6] Thus, while pharmaceutical companies contribute to this saturation, they are not its sole cause, and the appeal of psychotropic drugs is better understood as one consequence of a more general and far-reaching (and as a result, more insidious) cultural process of rhetorical saturation.

In other words, criticisms of corporate deception must be accompanied by accounts of rhetorical processes that condition individuals to brain-based ways of thinking and acting about the self. As Greenburg notes, pointing to shrewd corporate marketing strategies and Big Pharma's aggressive sales tactics "is no more or less illuminating than uncovering gambling in Casablanca."[7]

To explain the effects of popular neuroscience's cultural saturation, I take up a theme prefaced in previous chapters: brain-based ways of thinking are also entrepreneurial ways of thinking about life. In other words, there is an affinity between neurobiological paradigms and economic paradigms: conceptualizing life in terms of neurons, chemical processes, and connections makes it amenable to various economic-oriented calculations, including assessments of risk, accumulations of quantity ("health reserves"), and interventions rationalized by cost-benefit logics. Through the language of the brain, both health and happiness are made calculable, and the question "Am I not happy because I am sick?" is followed by an entrepreneurial one: "What investments can I make—actions, drug treatments, self-training—that will maximize my amount of happiness and health, and decrease my chances of illness and unhappiness?" As I will argue, the dispersal of brain-based, entrepreneurial discourses for speaking and acting condition the success of psychotropic drugs. Within a paradigm where happiness is viewed in terms of ideal quantities and activity levels of brain chemicals, the use of chemical interventions to finesse and improve these processes makes sense and appears to be a logical, humane, and efficient means of optimizing self and society (including one's children). Thus individuals are not so much brainwashed to accept pharmaceutical remedies for common distress as they are conditioned to understand life as a calculative project that they are responsible for assessing and quantifying through neurobiological lenses.

To explore the relationship between popular neuroscience and psychiatric culture (shorthand for the proliferation of psychotropic drugs, prescriptions, and related discourses, including advertising), I analyze a museum exhibit titled *Brain: The World Inside Your Head*. The exhibit opened at the Smithsonian in 2001 and continues to travel to a number of museums and health and science centers around the world.[8] My analysis is focused on the exhibit as it appeared at the Health Museum of Houston in fall 2007. The major sponsors of the exhibit are the National Institutes of Health, the federal agency responsible for coordinating health-related initiatives, and Pfizer, one of the world's largest pharmaceutical companies and the manufacturer of Zoloft, a popular antidepressant. *Brain* is produced by Evergreen Exhibitions, a professional company that specializes in exhibit marketing. Thus the exhibit is positioned in a complex and dynamic network composed of pharmaceutical companies, government agencies, educational and medical institutions, and families and individuals. In my analysis of *Brain*, I focus on how popular neuroscience disseminates neurochemical vocabularies and interventions through the complex rhetorical formation of a museum exhibit. The *Brain* exhibit blends education, entertainment, and marketing, and this discursive flexibility makes it a key hub for both the spread of popular neuroscience across diverse social institutions and the distribution of neuroscientific vocabularies to children and families. Thus, while the content of *Brain* illustrates key features of popular neuroscience, its formal features provide an example of how brain-based terminologies come to saturate popular culture. It is worth noting that, as with the other case studies of this book, my objective is not to challenge the scientific validity of chemical imbalance theories or the efficacy of psychotropic drugs. While many compelling critiques along these lines exist, I am interested in tracing and assessing the rhetorical processes of brain culture, not in judging its truth or falsity against some objective standard or absolute reality.[9] Thus, in the following analysis, I bracket questions of validity and focus instead on the ways in which neuroscientific languages come to saturate popular culture and facilitate brain-based ways of thinking and acting that include, but are by no means limited to, the consumption of psychiatric medications.

In the first part of this chapter, I situate *Brain* as a novel form of pharmaceutical advertising that operates through the space of the museum. Museums have increasingly become areas that blend entertainment and education, emphasizing fun, choice, and involvement as essential dimensions of the learning experience. In this context, museums are inviting hosts for marketing discourses because they encourage visitors to actively participate in exhibits and relate the museum experience to their own lives. After describing the rhetorical composition of *Brain*, I closely examine its unique features to better conceptualize how it functions as an agent of saturation. Because the exhibit inhabits a post-museum, Eilean Hooper-Greenhill's term for contemporary museum spaces that coordinate marketing, entertainment, and education, it is able to advance popular neuroscience's cultural saturation along both vertical and horizontal axes. By vertical saturation, I refer to the exhibit's ability to engage individual visitors, especially children, and persuade them to adopt brain-based ways of thinking about their own lives. By horizontal saturation, I refer to the exhibit's ability to spread brain talk across diverse social spaces. Because the post-museum is part of a network that connects corporations, educational institutions, government agencies, scientific research organizations, nonprofit interests, and families, its themes and concepts can move easily from one institutional space to another. Moreover, this circulation is facilitated by popular neuroscience's translatability, or its capacity to accommodate multiple meanings and many areas of life. I conclude by discussing the implications of Pfizer's covert marketing practices. While I am critical of Pfizer's role in *Brain*, I am convinced that a thorough assessment of psychiatric culture must go beyond criticism of corporate deception and examine the rhetorical factors contributing to the cultural saturation of brain talk.

Direct-to-Consumer Advertising and Neurochemical Selves

Psychiatric drugs are as much a part of popular culture as they are a medical phenomenon, as evidenced by the familiarity of name-brand drugs like Prozac, Ritalin, and Zoloft, and their frequent appearances in jokes, television shows, and movies, and on T-shirts, mugs, postcards, and even jewelry.[10] This cultural enthrallment with psychiatric medications is fed by the direct-to-consumer advertisements from pharmaceutical companies that litter television, magazines, and the pamphlets found in doctors' offices and sometimes schools. The dramatic increase in public messages about psychiatric drugs is positively correlated with an explosion of drug consumption. Between 1990 and 2000, sales of psychiatric drugs in the United States increased by 600 percent.[11] In 2007, Americans spent twenty-five billion dollars on antidepressants and antipsychotics, an amount Robert Whitaker notes is more than the gross domestic product of Cameroon, a country of eighteen million people.[12] Since 1997, when the FDA made major changes advancing the marketing of prescription pharmaceuticals directly to consumers, psychiatric drug advertising has exploded. Statistical and anecdotal evidence suggests that this marketing is effective, as more and more people are actively seeking out medical attention and asking for particular brand-name drugs. But pharmaceutical companies must sell more than a pill; because of the unique nature of their product (it cannot be purchased by consumers directly), the success of these companies depends on consumers motivated to pursue psychiatric diagnoses, and, more broadly, consumers accustomed to a certain way of thinking and acting with regard to the self. In order to purchase this product, a consumer must identify personal distress, recognize this distress as a potential disease rooted in biology, and then pursue medical attention in the form of a doctor's appointment and perhaps a direct request for a specific diagnosis and prescription. This readiness to identify one's personal distress as medically significant and to define one's life in the language of symptoms and disorders attests to a culturally dominant way of thinking that is biological and, more specifically, neurochemical. The idea that personal ailments are the result of chemical imbalances in the brain is a cornerstone of contemporary psychiatry, and this point of view is widely disseminated in pharmaceutical advertisements, a message epitomized by the ubiquitous sketches comparing the quantities of neurotransmitters between neurons in "healthy" and "depressed" (or otherwise afflicted) brains.

The incredible success of prescription pharmaceuticals marketed for psychiatric conditions is in part due to rhetorical processes that function at the level of the public grammar. Everyday individuals, or "lay" citizens with no particular medical or scientific expertise, are increasingly using neuroscientific terminologies to speak about themselves and their experiences.[13] Examples include the widespread circulation of expressions such as "my serotonin levels are low today" or "he must have a chemical imbalance." These colloquialisms exhibit a conception of life that is both neurobiological and economic, in the sense that life function can be articulated in terms of quantities, balances, and deficits. When individuals come to speak about themselves in these biological languages, they are primed to act on themselves through their biology to maximize quantities of health and minimize illness. Drug advertisements reinforce this sensibility by framing medications as products designed to "rebalance" brain activity and shore up brain connectivity to preserve life's delicate equilibrium. In the marketplace of products and practices that individuals can select to pursue health and prevent illness, drugs are uniquely privileged, in part because of their medico-scientific status in comparison with other options, such as meditation or journaling. While not all brain imaging studies support drug treatments, they do support the messages that all of life can be traced to biological processes and that the brain is an array of interconnected, dynamic processes that are both precarious and malleable (and can be tweaked for superior function). Both of these messages contribute to the brain-based paradigms that make drug treatments so attractive and commonsensical. Thus, in part because of brain imaging and the scientific authority it lends to popular neuroscience, drug treatments are privileged mechanisms of self-optimization. At the same time, however, drug treatments enjoy a special notoriety because their economic success adds credence to claims of corporate greed and manipulation, and also because their extraordinary abilities to affect mind and behavior raise significant questions about the authenticity of human identity (as seen, for example, in the uneasy attitudes expressed by the President's Council on Bioethics, discussed in chap. 1). In this chapter, I am primarily interested in accounting for the processes of cultural saturation that make drug treatments a privileged technology for self-fashioning, although it is important to note that drug treatments are not universally accepted and their popularity is contested.[14]

The Houston Health Museum: "We're All About You!"

The *Brain* exhibit is associated with a particular type of institutional space—the museum. As Eilean Hooper-Greenhill describes it, the museum is the "archetypal institutional form of the modern period," but its form and function have shifted in recent decades.[15] I adopt Hooper-Greenhill's concept of the post-museum to summarize the ways in which contemporary museums are increasingly modeled on businesses, assimilating marketing, consumer focus, and corporate sponsorship into their agendas. These tendencies toward an economic orientation have gained momentum over the past few decades and have been described by diverse theorists, including Timothy Luke, Sharon MacDonald, Tony Bennett, Barbara Kirshenblatt-Gimblett, and Donald Preziosi.[16]

Although these museum scholars differ in their precise character-izations of contemporary museums, there is a general concurrence that museums and science centers are increasingly immersed in marketing dis-courses—terminologies such as "customer," "niche market," "brand," and "corporate image" frame the mission of these post-institutions.[17] Museums have long been recognized as governmental institutions, "exhibitionary complexes" that shape individuals and collectivities.[18] Contemporary muse-ums, or post-museums, participate in a "commitment to dialogic and mul-tisensory forms of visitor engagement" that challenge "authoritarian" forms of didacticism.[19] Post-museums understand their visitors as active consum-ers who seek entertainment and participatory involvement. Contemporary museums still serve an educational function, as Luke explains, but one that is conceived along the lines of entertainment as "a completely packaged ensemble of amusing consumer choices," or what MacDonald describes as "supermarket logic."[20]

The Houston Health Museum is better characterized as an interactive science center rather than a science museum. Science centers feature open spaces that invite audiences to "play" with devices and "discover" principles for themselves.[21] They are less didactic and more participatory in their formal features. The Health Museum describes itself as "committed to excellence in innovative and interactive health and science educational experiences," and as "Houston's most interactive science learning center." The post-museum offers a variety of information and activities from which visitors can pick and choose, piecing together their own individualized "experience" from the available raw materials. The themes of individual agency and consumer empowerment are expressed on the Houston Health Museum's website, which features three smiling medical professionals, suited in scrubs and complete with stethoscope and professional markers, set against the motto "We're all about *you*!"[22]

The emphasis on individual choice is communicated in the composition and content of the museum space and the *Brain* exhibit. The spatial composition of the museum and the exhibit's interactive modules foster a sense of freedom and individual choice, promoting active engagement with the neuropsychiatric discourses presented. The emphasis on choice is apparent in the initial contact with the museum space. Upon entering the museum, a large entrance hall opens up into the featured exhibits. The enormous space in the center of the building invites visitors to choose their own path, to make their own choices about which of the "worlds" to enter.

The space of the *Brain* exhibit is also open: it consists of six thematic arrays (e.g., "The Living Brain" and "Brain Discovery"). Each station features a number of different modules that might include visual and verbal poster displays, video clips, actual objects, and interactive elements. None of the modules extends to the top of the space, adding to the impression of an unrestricted open space scattered with a range of media. In hallmark post-museum fashion, the exhibit is spread out over "a layout with no fixed direction or predetermined route," planned to "involve the visitor in making choices."[23]

The entrance to the exhibit is a canopy of giant purple "neurons" that arch over the visitor. The neurons have space between them, and there are montages of "axons" and "dendrites" that do not completely fill the space they occupy. The entrance is more of a suggestion—a sketched hallway more than an enclosed space that mandates the visitor's movement. Alternatives are available: it would be possible to enter the exhibit without encountering the purple neuron configuration, passing to the left or the right.

Pfizer, the exhibit's major sponsor, maintained for many years a webpage that featured a virtual tour of the *Brain* exhibit.[24] Virtual visitors are presented with a constellation map of the exhibit, and they can click on the links for each array to receive additional information about the featured theme. At the Houston museum, the space of the exhibit mirrors its virtual form but is even less linear. The virtual tour displays the map of the exhibit, illustrating the dispersed arrays with their diversity of modules. For the virtual tour, the six modules are numbered, suggesting a type of linear sequence to direct the visitor's perusal. In the actual exhibit there is no such sequence. The six arrays are spaced in such a fashion that it would be possible to visit them in almost any order, or to visit some while ignoring others. There are no arrows, numbers, or ordered maps that dictate the visitor's method of accessing the museum. The interconnected modules provide the appearance of choice, but they actually form a constrained labyrinth. Their arrangement in the compact space of the exhibit makes it likely that a visitor will encounter most of the modules, even if wandering without a specific agenda.

FIGURE 11 Visitors to the Houston Health Museum enter the *Brain: The World Inside Your Head* exhibit through a flashing canopy of "neurons."

Photo courtesy of Evergreen Exhibitions, *Brain: The World Inside Your Head* exhibition

Interactives

Popular neuroscience is a socially powerful language because it can refer to a wide range of social practices and events while at the same time connoting objectivity. In slightly different terms, brain-based vocabularies are authoritative "because they derive from the rational discourses of science, not the arbitrary values of politics."[25] These languages are not exclusive to scientists, however; individuals, too, are encouraged to understand and articulate their own experiences through these languages. *Brain* emphasizes its close ties to the medical profession and scientific research, thereby enhancing the exhibit's scientific ethos. This scientific resonance provides credibility and truth-value, but the exhibit is not a one-way authoritative address. Science itself is democratized, transformed into a participatory venture in which visitors are expected to "make scientific principles visible to themselves through the use of touch, smell, hearing, or the sense of physical effects on their own bodies."[26]

Brain: The World Inside Your Head features twenty-two interactives that, according to the exhibit's producers, invite visitors to "touch, grab, manipulate, stand on, sniff, listen, hold, pull—even walk all over—the exhibit, which

represents the latest research on the body's center of thought and information processing."[27] These interactives have two related purposes: First, they
engage their audience as active participants in the learning process—the
displays interpellate, or construct, visitors as consumer-agents responsible
for choosing to access the knowledge embedded in the exhibits. Second, the
interactives encourage visitors to interiorize this knowledge and apply it
to their lives, adopting the biological vocabularies to express and organize
their experiences. This adoption of ways of thinking and acting in relation
to the self is what Foucault describes as "technologies of the self," specific
ways of caring for the self that blend knowledge and practice.[28]

First, these modules encourage audiences, particularly young children,
to engage the exhibit through sight, sound, touch, and smell, turning the
passive museum experience of observing and reading into an active and
entertaining pursuit. Viewed as a marketing campaign, the interactives
are designed to elicit audience participation and increase the likelihood
that visitors will receive and retain the producers' messages.[29] Educational
research suggests that active learning strategies are more effective methods
of transmitting information, particularly in the sciences.[30] The National Science Education and Teaching Standards reflect this research and emphasize
the "doing" of science as a vital component of science education. In the
context of marketing, the exhibit is more likely to successfully distribute
psychiatric vocabularies than television advertising, for instance, which can
only engage audiences in a relatively passive role.

Second, the interactives are not only a productive formal feature
designed to enhance message effectiveness; they also instruct visitors how
to act and behave in relation to their own brains. Interactives, Andrew
Barry writes, have a political function, to "foster agency, experimentation,
and enterprise, thus enhancing the self-governing capacities of the citizen.
Interactivity promises, in other words, to turn the museum visitor into a
more active self."[31] The audiences are acquiring more than scientific knowledge—they are instructed in practical truths, or technologies of citizenship,
that have specific application to their lives. In a society where citizenship
is increasingly defined in terms of health, these technologies take the form
of illness prevention and health maximization, and include continuous
techniques of self-surveillance and self-management.[32] These technologies
are distributed as explicit exhortations to a particular practice of living, one
that habituates individuals to the use of biological categories for framing
daily life.

Many of the modules have clear applications to individual practices of
living. For example, "Boost Your Brain" is an activity that requires players to
spin a game wheel and keep track of the designated point values for each
option. Players gain points for activities that are healthy for the brain, but

points are detracted for "bad brain" activities. The virtual tour station states, "Don't leave the health of your brain to chance. You can boost your brain with these activities." Activities that promote brain health include nutrition and exercise, doing a crossword puzzle, cooking something from scratch, and taking a shower with one's eyes closed. Nearby modules offer detailed descriptions of the destructive activities, such as drug and alcohol abuse, in terms of their effects on brain biology.

The "Boost Your Brain" interactive illustrates two features typical of popularized psychiatric discourses. First, everyday activities, ranging from the essential to the mundane, are articulated in terms of brain biology. All are described as biologically relevant events that must be assessed in terms of their effects on neural processes. The value of an activity, from camping to drinking alcohol, is determined with reference to the brain. The brain becomes a meta-standard, the ground of judgment and the foundation of a vocabulary that can translate diverse occupations, from the necessary and universal (sleeping, eating) down to the more frivolous and particular (doing crosswords, starting a new hobby), onto a common plane.

Second, the exhibit posits a direct relationship between behaviors and the brain—individual activities directly affect the brain at the level of its biological functioning. The interactive is contextualized as a game because it involves chance. When an individual spins the game wheel, there is no way to control what activity is selected—bad brain or good brain. The game is framed, however, as instruction in a real-life correlation that leaves individuals in control—chance is eliminated in favor of individual responsibility. Each of the activities is a willful activity that individuals are encouraged to choose to engage in. Even sleep is described as at least partially a function of choice— brains need eight hours of sleep to retain their health, and individuals must take active steps to ensure that they accumulate this amount of sleep.

This emphasis on the individual's role in achieving and maintaining optimal brain function is what Monica Greco calls a "duty to be well," where health and illness are transformed into "vehicles for the self-production and exercise of subjectivities endowed with the faculties of choice and will."[33] Health becomes an enterprise of the self, regulated by a logic of capital: the goal is to simultaneously maximize health while minimizing risk. As described in previous chapters, the focus is not limited to alleviating exist- ing illnesses, but also to regulating potential illnesses through calculation, management, and strategic interventions. This shift entails a conversion in one's way of thinking, from viewing health as a normal state to viewing it as a quantitative resource that can always be increased, and as an achieve- ment infinitely postponed through the endless identification of new risks. The critical factor in this process, Greco writes, is "the positing of a personal susceptibility which is logically prior to cause."[34] In terms of the "Boost Your

Brain" interactive, the implication is that one can always do something to accumulate more health and to alter the risk calculation in one's individual situation: more activities of the beneficial variety will increase the likelihood of a healthy brain, whereas each activity in the destructive category alters the odds against brain health. Each individual is continuously responsible to see that "beneficial" brain activities outweigh "destructive" ones. Through the discourse of the brain, activities and experiences are rendered as calculable, measurable, and manageable. Further, the need for calculation and management is not confined to clearly marked states of pathology.

Even those modules without explicit "morals" or clear-cut applications to individual behavior function within these parameters. Visitors are interpellated as biological citizens, understanding their thoughts and behaviors as biologically correlated.[35] For example, "Brain Waves" invites visitors to view a simulated EEG readout of their own brain activity. EEG is a brain imaging technology that theoretically translates electrical signals produced by brain activity into graphic data that can localize the precise source of the activity. For instance, if an individual thinks sad thoughts while undergoing an EEG, the data will show which region of the brain is responsible for the production of those thoughts.

In "Brain Waves," visitors "lean on electrodes and perform tasks to see real-time EEG measurements and simulated imaging of corresponding brain activity." The interactive features a stand with a padded bar on which individuals are to place their foreheads. A computer is positioned in front of the visitor. The screen features the simulated EEG readout in the form of a dynamic graph, which is surrounded by coded information. For example, the far-right column of the screen lists a series of terms—"Gain," "Sweep," "Delay," and "TRG Level"—and numbers are listed below each term. There is no visible key to decode the terms or the abbreviations. The graph itself features several wavy lines moving horizontally across the screen. Regardless of what task the visitor engages in—thoughts, movements, or speech—the graph produces these dynamic and variable lines. The dynamism of the graph implies that each movement of thought or of the body produces a different set of data, allowing the EEG readout to precisely correlate every activity with internal, biological processes. The graph itself is not legible—there is no way for a layperson to decipher what the various terms, codes, and data signify—but it effectively communicates the idea that everything an individual does has a precise correlation in the brain, and that science can access, visualize, and calculate this information. In fact, the illegibility possibly increases the likelihood that the graph will be consumed as veridical scientific data.[36]

A similar interactive, "Back and Forth," contains three stations that allow visitors to visualize the brain activity that corresponds to the reflexive experience of pulling back from a hot stove, automatic functions such as breathing, and balancing for as long as possible. Each aspect of life is, at its most basic level, a biological event with precise neurochemical correlates. Another simple interactive invites visitors to complete a puzzle, matching specific parts of the brain to a diagram labeled with different functions such as "Vision," "Touch," and "Personality/Emotions." The brain is depicted as a collection of discrete parts, or regions, each responsible for a specific part of an individual's life. The interactive communicates more than abstract knowledge—it instructs visitors in a materialist-neuroscientific paradigm that is a mode of understanding the self as a fundamentally biological being. However an event might appear on the surface of consciousness and awareness, it is always biological and is produced in the interior of the body.

While the brain puzzle suggests a relatively simple structure-to-function relationship reminiscent of older localization theories, many of *Brain*'s rhetorical features communicate a more dynamic, interconnected, and plastic brain. These elements of the exhibit frame life as a balance that must be maintained, an assembly of distributed chemical and electrical processes that require constant calibration. For instance, an interactive called "Synapse Pop!" features a mural of enlarged axons and dendrites. In front of the

FIGURE 12 This exhibit encourages visitors to learn which parts of the brain are responsible for which functions by putting together a simple puzzle.

Photo courtesy of Evergreen Exhibitions, *Brain: The World Inside Your Head* exhibition

mural, a series of purple tubes (neurons) are lined on a track, and when visitors place a ball (representative of a neurotransmitter molecule) on the track, it flies along the track, apparently jumping from neuron to neuron across the synaptic gaps, setting off flashing lights as it speeds throughout the brain's circuits. The exhibit is designed to illustrate the electrical and chemical activity that constitutes brain function. Accompanying text included on Evergreen's website states, "The message is clear: Your brain is exciting. Your brain is surprising. Your brain is always changing. . . . All brain function, everything we are and do, begins with neurons and synapses, electricity and chemistry."[37] The interactive suggests that life ("everything we are and do") is most authentically understood at the level of neurons and their electrical and chemical processes. Moreover, the relationship between the neurotransmitter (ball) and the "exciting" flashing lights dramatizing brain function drives home the message that brain activity is composed of minute biochemical processes, and that proper function depends on mobile "communicating" neurotransmitters. Without the work of the neurotransmitter, the brain system would not light up and exhibit its exciting, dynamic activity. Thus "Synapse Pop!" instructs visitors in a chemical understanding of brain function, one that resonates with advertising discourses situating drugs as mechanisms of balancing and regulating chemical activity. Through these dramatizations of brain process, life itself takes on a sense of precariousness and contingency that reinforces the urgency associated with popular neuroscience's cultural saturation. If life comprises minute chemical processes, then anything might alter the brain's delicate chemical balance and impair function.

Other highlights of the exhibit drive home the message that the brain is a space of processes requiring ongoing calibration. A graphic display headlined "Seeing Is Believing" depicts a Dali-esque landscape featuring a human head crosscut with a "lighting up" network similar to the one in the Nintendo advertisement discussed in chapter 1. A miniature man stands atop the head, fully clothed in a business suit. Although the graphic is amenable to diverse interpretations, the accompanying text suggests that it is designed to illustrate the idea that brain connectivity is responsible for producing the "whole person," represented by the clothed man standing on top of the detached head. The text questions visitors, "Is what you see real?" and responds with a definitive "No!" Sight does not allow direct access to the outside world, but rather is mediated by the brain: "You don't see anything directly, but only after your brain processes its electrical signals." The text concludes with the reminder that "your brain creates your reality." The graphic reinforces the familiar message of popular neuroscience, that one's identity and reality are produced by brain activity, and it also extends this

FIGURE 13 The Synapse Pop! exhibit allows visitors to watch a small "neurotransmitter" ball as it courses through the brain's circuits, accompanied by flashing lights, to illustrate the brain's electrical activity.

Photo courtesy of Evergreen Exhibitions, *Brain: The World Inside Your Head* exhibition

message by suggesting that the primary determiner of one's reality (perception of self and one's world) is more closely related to brain activity than any external referent or "objective" world. In other words, reality is not the product of an encounter with an external world, but instead is manufactured by the brain and, as such, caused by the dynamic electrical and chemical activities that course through the space of the brain. In the context of the exhibit and brain culture more generally, this message supports themes of optimization, choice, and control: by changing one's brain, one can literally create one's world anew. Becoming happy or peaceful or successful depends on manipulating one's biology, not on altering external social and economic circumstances. The flip side of this message is, of course, that suboptimal brain activity can produce an undesirable reality, creating a world characterized by negative affect, social and economic failure, and general malaise and distress. As I will explain, it is in the context of this overall message—that the brain (and not external circumstance) is the determiner of reality—that pharmaceutical drugs acquire such allure. If the self and its reality are

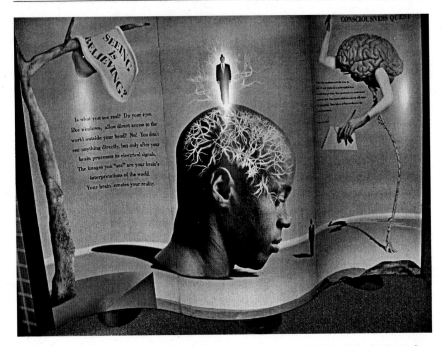

FIGURE 14 This display suggests that the world is but a creation of the brain's electrical and chemical activity, with no genuine correspondence to any external reality.

Photo courtesy of Evergreen Exhibitions, *Brain: The World Inside Your Head* exhibition

products of brain chemicals, chemical interventions are efficient and attractive means of controlling and shaping both self and world.

Democratizing Scientific Language

The active learning techniques and the instruction in technologies of the self constitute a biological discourse that exhibits a strange causality. Experiences—behaviors and emotions—are articulated as events that are simultaneously determined by interior biological processes and are the causes of these processes. Experiences are the result of the brain, and they are also the means by which the brain is fashioned into a healthy and productive entity. The result is a discourse that identifies individuals with their brains (the exhibit frequently states, "You are your brain") and promotes a constant vigilance. Each experienced emotion and behavior becomes an opportunity to assess the health of the brain—a "symptom" to reflect on and interpret. At the same time, behaviors become opportunities to shape the brain and improve its health. Discourse about all things—whether experienced as willed behaviors or passive experiences—must "pass through" the brain and find expression in the biological language of contemporary neuroscience.[38] These vocabularies promote an attention to and a practice of the self, expressed in self-diagnosis and self-monitoring, all centered on the elusive health of the brain.

It is through the mediation of neuroscientific vocabularies that subjective experiences, including emotions and ideas, become capable of rational management. They are made amenable to capital as resources or values that can be put to use. Experiences are not only something each individual "has," but also are scientifically valid "symptoms" that can be assessed, manipulated, and capitalized for optimal health. Individuals are encouraged to adopt these scientifically valid languages as expressive of their personal experience. For instance, the interactives highlight the electrical and chemical composition of the brain—the theme "your brain is a lightning storm" appears frequently, visualized by images of the brain with flashing "lightning bolts" coursing across its terrain. In the neuron canopy entrance, strobe lights mimic the electrical activity of the brain as a video display intones, "Your brain processes your thoughts, moods, and emotions. Everything you do and everything you are comes from your brain. Everything your brain does comes from electrical signals and chemical connections. At its core your brain is a lightning storm." Similarly, the "Brain Waves" exhibit features this image of the brain coursed by lightning bolts, suggesting that when imaging technologies visualize brain activity they are accessing this electrical and chemical activity.

Through these interactives, the highs and lows of mood are articulated as equivalent to the highs and lows of chemical quantities and the strength and quantity of electrical signals. Feeling, or subjective experience, becomes a marker for assessing brain function. In other words, "I feel sad" is rendered equivalent to "my serotonin is low" or "my brain activity in my amygdala is low." Feelings and mood are not simply signs or markers of a related but distant biological condition; they *are* the biological condition. The process of translation renders them equivalent, two ways of expressing the same thing. Feeling sad *is* low serotonin levels. Both forms are equally expressive of a subjective state, mood, or feeling. Within the context of this translation, scientific language has the capacity to communicate personal, subjective experiences. Experience then becomes a way of constantly assessing the health of the brain—each experience expresses something about brain function, not as a symptom but as a direct manifestation. Subjective vocabularies ("I feel sad") and scientific vocabularies ("my serotonin levels are low") are parallel expressions of equivalent phenomena.

By way of rhetorical formations like *Brain*, neuroscientific terminologies saturate popular culture as ways of communicating subjective experience. This saturation makes it more likely that individuals will pursue biochemical interventions, even absent direct manipulation or deception by corporations. For instance, when the experience of sad feelings is understood as a "chemical imbalance," it is more likely that the afflicted individual will pursue chemical remedies (namely, psychotropic drugs).[39] In the neuroscientific rendering, sad feelings acquire new social utility: they are symptoms that communicate both certain facts about an individual's health reserves and recommend particular interventions.

Marketing, Educating, Entertaining

Earlier in this chapter, I described saturation as a process with two trajectories: a horizontal process in which popular neuroscience spreads across different social spaces, and a vertical process whereby popular neuroscience is disseminated to individuals for their active uptake and personal usage. While this picture of saturation is useful to an extent, in reality the rhetorical movements of popular neuroscience are far more complex and interconnected than those suggested by the simplified image of a grid with horizontal and vertical axes. The diffusion of popular neuroscience is better conceptualized as a discursive network of mobile coordinates and dynamic relays. *Brain* is a case in point: the exhibit, created through the collaboration of scientists, corporations, and government agencies, brings teachers and students into the museum space and works to assimilate them to neurobiological terminologies. These terminologies travel with teachers and students to the classroom, where they are further reinforced through exercises and activities. When students return to their families, they take with them their homework and their stories, ferrying popular neuroscience to other social spaces. Brain-based rhetoric moves easily among these different spaces and individuals because it is highly translatable, as seen, for instance, in the equivocal languages of health that can activate moral, biological, and social meanings. In part because of this discursive flexibility, ways of thinking about the self that are learned in the museum can be useful in the home, at school, or in other social venues. In the remainder of this section, I explore how the cultural saturation of popular neuroscience is caught up in broader social trends involving the dissemination of economic- and market-based models of personal and social governance.

The mission of the post-museum blends marketing and education, making it attractive to corporations whose business intersects with some dimension of science. Evergreen Exhibits, the producers of the *Brain* exhibit, frames their service as a novel marketing strategy. Using the museum as a medium, their website states, will reach "large, qualified, captive audiences" with "credibility, a degree of acceptance, and a level of direct consumer involvement unmatched in the world of sponsorship." Exhibit marketing promises a return fifteen to twenty times the initial investment by providing a means of "extending the corporate sponsor's brand with consumers."[40] Exhibit marketing is thus an attractive option for pharmaceutical corporations such as Pfizer, which is increasingly turning to educational and advocacy forums to distribute favorable messages. These campaigns rarely discuss the product directly (the *Brain* exhibit does not mention Zoloft); instead, they work by conditioning audiences to vocabularies and modes of understanding that are congruent with seeking out pharmaceutical products.

Despite financial sponsorship, the *Brain* exhibit cannot be viewed as a simple production of Pfizer. Participants include Evergreen Exhibitions, a private company that specializes in exhibit design and boasts of a number of Fortune 500 clients, and the National Institutes of Health, the primary federal agency managing medical research. In addition to these producers, different agents sponsor the *Brain* exhibit at each of its local appearances.

The *Brain* exhibit overspills its contingent spatial and temporal boundaries, functioning as a "contact zone" that brings multiple actors into interaction and extends the reach of psychiatric vocabularies to permeate locations external to the museum space.[41] The traveling exhibit proper is only one aspect of the exhibit experience, as a nexus of events and activities circulate around the exhibit. These events range from lectures to entertaining celebrations of the brain, and each involves its own miniature network of sponsors and agents. The Houston exhibit, for example, is accompanied by the "Brain Basics Speaker Series," which features medical professionals from local hospitals and research institutions addressing topics ranging from meditation to brain geography. In addition to the lectures, the museum features a special week of spring break activities for children and families centered on the theme of the brain. The events include "brainy crafts" and a "Brain Fun Night," which invites visitors to "find out cool stuff" about their brains and participate in exciting activities.[42] The Dana Foundation and the Neuroscience Research Center at the University of Texas Health Science Center sponsor the Brain Fun Night. Other activities include as sponsors the Baylor College of Medicine, AstraZeneca Pharmaceuticals, and Shell Oil Company.

Because the exhibit is part of a science museum, it is able to extend psychiatric discourses into schools, homes, and communities. For example, the exhibit is marketed as a desirable destination for school field trips. The Houston Health Museum features "No Brainer," which is billed as an "Excellent Field Trip Opportunity." The "sensational spring field trip" is described as a two-hour event that includes a tour of the exhibit, a sheep brain dissection, and an interactive science class in which educators answer the question of "how your brain really works." The Houston museum estimates that at least forty-five thousand schoolchildren visit its special exhibits each year as a part of school-sponsored field trips. To augment these field trips, Pfizer distributes a special guide to educators that includes a variety of classroom lessons and activities designed to encourage discussion of and attention to the brain.

Field trips are increasingly viewed as productive marketing schemes for a variety of industries. Such activities are desirable mediums because studies suggest that the active experiences result in a high level of information retention by students, and industry-related field trips have been shown to increase retail sales by up to 18 percent for the targeted industry.[43] Field trips are not just one-time events; rather, they are viewed as contact points for establishing a continuum of contacts with the relevant institutions that extends "well beyond school and childhood."[44] In the case of science education, field trips are viewed as a productive pedagogical model primarily because students actively engage the material from a variety of perspectives.[45] Because science museums identify themselves with an educational mission, they are common field-trip designations. Sponsoring corporations are guaranteed a captive audience and a pedagogical framework that extends the messages of the museum space into the space of the school and home. By targeting children as a primary audience, the exhibit is designed to reach a broader audience, including parents, families, teachers, and school administrators. Schoolchildren are positioned as a relay point in this circulation system, agents who will disseminate the messages received through the exhibit to members of their school, family, and social networks.

In conjunction with the alliances between schools and corporations that are welded in the common space of the museum, pamphlets designed for educators and families encourage visitors to adopt a vigilant attitude toward brain assessment in daily life, beyond interaction with particular authorities or within particular locations. Pfizer distributes a *Teacher's Activity Guide* to accompany the *Brain* exhibit.[46] The guide contains a number of activities suitable for the classroom, designed to further students' educational experiences through active engagement with the psychiatric discourses. The activities, categorized as pre- or post-visit, share a general pattern: they encourage individuals to document their lives in the terminologies of the brain sciences. Many of these activities take the form of writing, requiring students to chronicle their lives in terms of their brains. The journal assignments cover a range of activities, including sleeping, eating, interacting with others, studying, and playing. These assignments extend the adoption of psychiatric discourses into virtually every sphere of life. Pfizer also distributes other literature at the exhibit and through its website that is similarly designed to disperse the discourses of brain biology across diverse social contexts. Parents, for example, are targeted with *Talking to Kids about Brain-Related Conditions.*[47] This literature does not directly mention Pfizer products—the pamphlets simply encourage the individual production of neuroscientific discourses. In addition, they promote Pfizer's brand, directing consumers to additional resources—such as Pfizer's website—that offer specific information about pharmaceutical products such as Zoloft.

Conclusion

Brain is undoubtedly a covert marketing tactic, but it does much more than sell Zoloft and other psychotropic drugs. It disseminates a lexicon of neurobiological terms that carries a way of thinking and acting that is both healthist and entrepreneurial. This lexicon undoubtedly conditions the success of psychiatric medications, but increasing the sales of psychotropics may not be its most important effect. Through rhetorical forms like the post-museum, neurobiological vocabularies and interventions are distributed to individuals as empowering tools that they can take up for themselves to use in maximizing health and life potential. *Brain* does not merely manipulate consumers into purchasing antidepressants and related medications—it instructs individuals in an entrepreneurial rationality, framing life as a business that can be run according to economic logics of investment, assessment, risk, and reward. The languages of the brain, and especially the languages of chemical imbalance, make this entrepreneurial rationality concrete by translating life into quantifiable, visible processes that carry the weight of scientific authority. At the same time, these languages accommodate notions of personal choice, empowerment, and active participation in learning and life management. Critical approaches that charge corporations with deceiving and coercively manipulating passive consumers often neglect the fact that psychiatric culture works through, and not against, personal desires and choices. These desires are shaped and constrained by powerful discourses, but they are not simply repressed or wholly invented by corporations or other controlling entities. In the next chapter, I explore more thoroughly the extent to which popular neuroscience accommodates the attractive languages of rights, freedom, and empowerment, and I reflect on the critical challenges posed by these rhetorical patterns. One theme of the next chapter is that the pro-freedom, pro-autonomy rhetorics of popular neuroscience situate treatments, including psychotropic drugs, as a "right" and a "choice," making it difficult for charges of corporate oppression to gain critical traction.

6

Mental Health Care, the Rhetoric of Recovery, and Entrepreneurial Lives

In 2005, A. Kathryn Power, director of the Center for Mental Health Services, a part of the Substance Abuse and Mental Health Services Administration (SAMHSA), enthusiastically announced, "We are in the recovery business!" Power described "a monumental societal shift in the way we understand and approach mental health and mental illnesses in this country," something "truly momentous" affecting every level of society.[1] Changing conceptions of mental health and illness are, as previous chapters have suggested, central features of brain culture. Definitions of mental health and illness have expanded to the point where virtually any aspect of life (job success, relationships, parenting, and so forth) can be located under the purview of mental health, in part because of persuasive brain images that can show the neurological basis of any mood, thought, or behavior. In this chapter, I focus on how these expanding definitions, conditioned by the scientific and cultural authority of neuroscience and brain imaging, influence public policy and affect the ways that individuals organize their lives in relation to social services. Specifically, I take up Power's notion that recovery is a business, and I attend to the ways in which the rhetoric of recovery is used to justify a market-based arrangement of social services on both scientific and moral grounds. This chapter extends a central theme of previous chapters: that rhetorics of neuroscience and mental health are key means of disseminating entrepreneurial models for living that prioritize self-optimization and risk management in the name of health and personal fulfillment.

Self-Directed Care and the Rhetoric of Recovery

In her speech, Power singled out Florida's Self-Directed Care (SDC) program as a role model of the "recovery business" to which other states and communities should aspire. The Florida SDC program, founded in 2000, is premised on a commitment to the related ideas that health care is intricately linked to economics, and that individuals' mental health is realized through their active participation in market-based activities (including consumption and production). The Florida SDC website explains that the program is based on the understanding "that individuals choosing services and making purchases will help them begin, or remain on the road to recovery." Florida SDC makes this understanding concrete by putting federal health care funds in the control of individuals—or, in the preferred terminology of mental health, "consumers"—with mental health problems. Consumers can select from a wide range of programs and piece together the plan of care, or "recovery path," of their choosing. Programs include traditional medical treatments as well as multiple varieties of alternative therapies (play, couples, etc.), work-related services, and even shoes and clothing. By engaging in active processes of choice and consumption, individuals can pursue, the SDC website states, "achievement of the highest level of desired personal wellness."[2] Each individual's pursuit of wellness is measured by a number of standards, including self-reports, the number of productive days in the community, and assessments by friends, family members, and "recovery coaches." The Florida SDC program is lauded because it is economically efficient and scientifically validated (relying on "evidence-based" practices), and because it incorporates the democratic values of autonomy, choice, and respect for individual diversity.

The Florida SDC program illustrates the extent to which conceptions of mental health are entangled with economics at the level of public policy. Florida SDC operationalizes health in economic terms: the healthy (or recovering) individual is a consumer who actively engages in market activities in order to optimize productivity and fulfillment. SAMHSA drives home this message in the 2005 report *Free to Choose: Transforming Behavioral Health Care to Self-Direction*: "Enabling individuals with disabilities to become active in the marketplace by encouraging them to choose and pay for their care represents another means of integrating them into the larger society," helping them to recover by allowing them to "become true 'consumers.'"[3] In the context of these mental health discourses, to be healthy is to be a consumer who exercises autonomy and control over one's life. The consumer identity also indicates an entrepreneurial identity in which consumption is itself a form of production—consumption, in other words, is valued because it is a way for individuals to maximize their productivity, happiness, and "human capital." This identity is not solely economic: the consumer role is also privileged as a moral, humane term that is a desirable alternative to the passive, dehumanizing "patient" identity.

The sweeping changes in mental health care policy exemplified by Florida SDC solder health to entrepreneurial models of personhood through the language of recovery. Recovery has become the expected language for speaking of mental health at all levels of government and society: it has been described as "the new buzzword," and "the most significant force for change in mental health practice since the closure of the hospital asylum."[4] Moreover, recovery is heralded as a rhetorical revolution: its connotative flexibility, Power states, "makes it possible for everyone to interpret, embrace, and apply . . . with a shared understanding."[5] Across state, scientific, and advocacy contexts, recovery is constantly being defined and celebrated, often in the same utterance, as if the very word exudes a magnetic attraction that compels people to ceaselessly redefine and praise it. In this chapter, I analyze the rhetoric of recovery, focusing on the President's New Freedom Commission on Mental Health, orchestrated by President George W. Bush. The commission's 2003 report, *Achieving the Promise: Transforming Mental Health Care in America*, is a rhetorical centerpiece of the recovery movement, and it continues to influence mental health care policy even after the conclusion of Bush's presidency. Through an analysis of *Achieving the Promise* and its broader socio-rhetorical context, I show how recovery's ability to activate diverse meanings—moral, ethical, and scientific—helps to frame life as a business in which individuals are responsible for maximizing their own capital. More specifically, I am interested in the policy consequences of these entrepreneurial models—in concrete terms, how do diverse understandings of mental health and illness, supported by brain imaging, affect public policy and facilitate various neoliberal patterns of social and economic organization? By neoliberal patterns, I am referring to programs that reduce state services, rely on market-based models of personal and social governance, and place increasing responsibility for risk management and maximization of economic productivity on individuals. In the case of *Achieving the Promise* and programs like Florida SDC, the scientifically authorized languages of recovery ultimately function to frame these types of programs as morally desirable, democratic, and empowering alternatives to state control. What is fascinating about this rhetorical process is that languages historically associated with social advocacy movements protesting state policies are now actively adopted by the state. In the remainder of this section, I explore this rhetorical process by describing the New Freedom Commission and situating it in the context of deinstitutionalization initiatives that have accelerated in the past four decades. In the next section of this chapter, I continue to examine how the New Freedom Commission's rhetoric of recovery unites both social advocacy discourses of democracy, freedom, and empowerment, and scientific discourses of neuroscience, brain imaging, and evidence. I conclude by tracing the ways in which the hybrid discourse of recovery justifies policy reforms that bring about models of living that situate individuals as entre-preneurs of themselves.

The New Freedom Commission and Achieving the Promise

On April 29, 2002, President George W. Bush announced the creation of the New Freedom Commission, a task force appointed to comprehensively review the state of mental health care in the United States.[6] This was only the second such commission in American history, following Jimmy Carter's presidential mental health commission more than twenty-five years earlier. In his introductory speech, Bush identified three major obstacles plaguing the current mental health care system: stigma resulting from misunderstand-ing, fear, and ignorance; fragmentation of services across federal, state, and local providers; and a lack of equity in coverage and reimbursement schemes compared to physical illnesses. By comprehensively addressing these "bar-riers" and providing concrete recommendations for policy solutions, the commission would, in Bush's vision, "work for a welcoming society, a society where no American is dismissed, and no American is forgotten."[7]

One year later, the commission released its comprehensive report, *Achieving the Promise: Transforming Mental Health Care in America*. The bleak document bemoans the "fragmentation and gaps" that plague a service delivery system marked by unfairness and discrimination. Too many Ameri-cans, the report states, "fall through the cracks" of the system and do not have access to the care and treatment supports vital to "recovery." The com-mission adopted the phrase "transformation" to signify that mere reforms would be inadequate to remedy these far-reaching problems. The scope of this transformation is articulated with six policy goals, with each goal includ-ing several specific policy recommendations deemed vital to its realization. The final report succinctly condenses the commission's agenda into two guiding principles paramount for successful transformation of the mental health system. First, services and treatments must be "consumer-centered" and "consumer-directed," recognizing each individual's unique needs and offering "real and meaningful choices." Second, mental health care must not limit itself to "managing symptoms" of identifiable illnesses. Instead, it must take a broader view of what constitutes mental "health," addressing "consumers' ability to successfully cope with life's challenges, on facilitat-ing recovery, and on building resilience, not just on managing symptoms."[8]

Although the commission was not the first voice to highlight recovery, it was a central means of cementing recovery's status as the privileged vocabulary for mental health care practice. Indeed, the commission is often praised for its rhetorical contribution—disseminating the language of recov-ery—rather than any particular policy change. Michael Hogan, chair of the commission, states, "Throughout the Commission's deliberations, the theme of recovery kept emerging with multiple meanings and implications."[9] In this chapter, I treat recovery as a rhetorical formation. The most prominent feature of this formation is, naturally, the word "recovery," which includes the definitional and epideictic activity that builds up around this word as it circulates across a number of cultural, medical, and public policy contexts. Three recurring terms that commonly cluster around recovery are "community," "choice," and "evidence-based." As these examples suggest, recovery is a powerful language because, like other healthist lexicons, it accommodates scientific and moral meanings. Although this chapter focuses on the New Freedom Commission and *Achieving the Promise*, it is important to note that the rhetoric of recovery is a much broader and far-reaching discourse and is not a specific production of the Bush administration. Recovery rhetoric was prominent in Carter's commission and its report, and it also circulated prominently in Clinton-era mental health discourse. More recently, the Obama administration has picked up recovery as the guiding discourse for mental health care policy, and to date the White House continues to operate within the rhetorical and policy parameters reflected in *Achieving the Promise*. The continuity of recovery rhetoric across different administrations suggests that mental health discourse is a prominent component of the neoliberal agenda that saturates contemporary politics and culture, and therefore is not limited to a particular party or administration.

New Freedom in the Context of the Deinstitutionalization Movement

A major theme of the New Freedom Commission report is that recovery from mental illness can take place only in the context of one's community. The report creates a stark contrast between the space of the community and the space of the institution, drawing the former as one of choice, freedom, and individuality and sketching the latter as a dark cavern of confinement, isolation, and coercive homogeneity. The New Freedom Commission is not the first federal initiative to trumpet community alternatives to institutional care. Over the past decade, the federal government has reinvigorated a discourse relating to mental health and illness that had waned since the Carter administration. This increased attention to mental health is witnessed in the first White House Conference on Mental Health, sponsored by Tipper Gore in 1999; federal consideration of important mental health–related policy initiatives, such as the Mental Health Parity Act, which demands equal reimbursement procedures from insurance companies for mental and physical illnesses; the increased number of official reports and documents, including the first surgeon general's report on mental health and mental illness in 1999, which catapulted the moral language of rights, diversity, equality, and empowerment into the mainstream mental health lexicon; and the increasing role of federal agencies devoted to mental health and illness (most notably SAMHSA and the NIMH).

Historically, the trope of recovery traces the path of deinstitutionalization and has emerged as the favored discourse for articulating various challenges to institutional care. David Satcher, in the 1999 surgeon general's report, locates the origin of recovery discourses in consumer-driven reform movements working for community care as a humane and effective alternative to institutionalization.[10] In part because mental institutions came to be seen as little more than warehouses for the aged, senile, and severely incapacitated, recovery became popular as shorthand for the hope of cure or, at the least, reintegration into normal community life despite disability.

The two fundamental principles for transformation identified by the New Freedom Commission—consumer choice and a broad understanding of recovery—are an essential part of the report's articulation of a pressing need to shift from institutional to community-based mental health care. The report insists that "the Nation must replace unnecessary institutional care with efficient, effective community services that people can count on," providing care in "integrated" settings instead of segregating consumers in institutional spaces.[11] The principles of choice and recovery can take place only in the context of the community, conceived as a space in which individuals can act out their choices and pursue fulfillment and empowerment. The New Freedom Commission on Mental Health was just one part of Bush's New Freedom Initiative, an agenda for promoting community care for all persons with disabilities. The New Freedom Initiative is designed to implement the landmark 1999 Supreme Court ruling *Olmstead v. L. C./E. W.* In the ruling, the majority opinion condemned the "unnecessary segregation" of persons with mental disabilities, and ruled that care must occur in the most integrated settings possible, affirming the "rights" of individuals to participate fully in community life.[12] The civil rights terms "integration," "segregation," "discrimination," and "rights" percolate throughout contemporary mental health discourses and permeate *Achieving the Promise*, which adopts the language of *Olmstead*, framing participation in the "community of one's choice" as both a "right" and a vital component of recovery.

Olmstead is indicative of a widespread antipathy toward institutional care for persons with disabilities that can be traced back decades. Historians identify the 1950s as a critical decade in the history of mental illness, widely recognized as a turning point in public and political sentiment toward mental health care.[13] The community care movement, comprising grassroots activists, former patients, and families, reached a critical threshold as their demands began to gain recognition from state agencies concerned about the economic inefficiency of institutional care. In addition, the development of the psychiatric profession as a medico-scientific entity contributed to the swelling support of deinstitutionalization, as psychiatrists saw the integration of mental health care into mainstream medicine as a source of professional credibility. Around this time, the introduction of apparently effective psychotropic drugs spurred the transition to community care because they helped many consumers to function in integrated settings, heightening the perception that institutional care was coercive and unnecessary. There are many historical accounts of the deinstitutionalization movement, and each recognizes a different combination of factors as causing institutions' fall from grace and the growth of support for community care. Some attribute deinstitutionalization to progressive humanitarian impulses, while Marxist scholars point to the economic necessities of rechanneling patients into the workforce. Others identify psychotropic drugs as the crucial catalyst, while some situate deinstitutionalization within a broader context of changes in welfare capitalism and the rise of neoliberal government.[14]

Whatever the diverse motivations involved, the alignment of government, professional, and advocacy interests in opposition to mental health care institutions resulted in several legislative policy initiatives designed to promote community care as the favored alternative. The strongest outpouring of criticism directed at institutional care fomented between the 1950s and 1970s, and extended from patient-focused grassroots movements to the upper echelons of the federal government. In 1949, the National Institute of Mental Health was created, with the explicit purpose of bringing about "the demise of public hospitals and to substitute in their place a community-oriented policy."[15] Shortly after its inception, the first head of the NIMH, Robert Felix, addressed the American Psychiatric Association, saying, "The guiding philosophy which permeates the activities of the National Institute of Mental Health is that prevention of mental illness, and the production of positive mental health, is an attainable goal."[16] Felix proved a strong advocate for recovery-oriented, community care reforms, capitalizing on an "awkward marriage" between fiscal conservatives and social reformers, who both saw their interests being served by deinstitutionalization.[17]

The deinstitutionalization movement reached a critical threshold when John F. Kennedy signed the 1963 Community Mental Health Centers Construction Act, a law that "culminated nearly two decades of ferment."[18] The legislation provided federal subsidies for community mental health centers and encouraged the continued dismantling of public institutions for the care of the mentally ill. The mental health historian Gerald Grob argues that Kennedy's legislation had mixed results.[19] Community solutions proved effective only for persons with "milder" afflictions, and those with severe disabilities found themselves without adequate services in many cases. Further, the legislation provided no means for coordinating federal, state, and local care, promoting "deep and bitter divisions" over program coordination and scarce financial resources. Far more influential in terms of deinstitutionalization trends, Grob argues, were the 1965 federal entitlement laws establishing Medicare and Medicaid. These programs encouraged states to transfer patients from state-funded public institutions to nursing homes and general hospitals, where the federal government would pick up the bill. The result was less deinstitutionalization in favor of renewed community integration, and often a "massive transinstitutionalization" motivated by financial considerations and enabled by a lack of federal, state, and local cooperation. By the early 1970s, policy solutions designed to reduce public institutions had resulted in "a confusing array of organized and unorganized settings for treating persons with mental illnesses."[20]

Carter initiated the first presidential commission on mental health in 1977 to address the apparent failure of government-promoted deinstitutionalization. Grob details the tumultuous events of the commission, which eventually resulted in the release of a report that informed the passage of the Mental Health Systems Act of 1980. The act continued the focus on community care solutions, and again provided grants to community mental health care centers and other agencies whose work fell under the purview of community-based mental health care. In an attempt to address the shortcomings of the Kennedy administration attempts, Carter's law mandated "performance contracts" and accountability measures to ensure some level of coordination among the diverse agencies responsible for implementing the federal mandates. Shortly after the passage of the law, Ronald Reagan was inaugurated and he quickly reversed Carter's policies, turning mental health care over to the states and dramatically reducing federal support.[21]

Rationalizing Neoliberal Reforms: Recovery's Fusion of Morality and Science

In the wake of these failed deinstitutionalization initiatives, one of the key rhetorical problems that the Bush commission faced was advocating for deinstitutionalization while avoiding the perception that its reforms were simply a case of the federal government encouraging the elimination of institutions without providing adequate community care as replacement, leaving individuals to "fall through the cracks." The New Freedom Commission addressed this problem directly, indicting the "fragmented" system and calling for one that was inclusive, "seamless," and "wraparound." Notably, despite its calls for comprehensive transformation, the commission did not call for any new funding in either *Achieving the Promise* or the subsequent *Federal Action Agenda* designed to implement the initial report's recommendations.

It might be more precise to say that the rhetorical problem was articulating transformation as a comprehensive, even radical, change from the status quo that, despite its scope, would not require additional resources, and would be likely to demand fewer federal resources than are currently allocated.

Recovery as a Social Advocacy Rhetoric

After its release, *Achieving the Promise* was lauded by policy and advocacy organizations alike for its landmark articulation of the "recovery perspective." The nonprofit Bazelon Center commended the report, announcing, "This is the first time that such a goal [recovery] has been articulated in any federal policy document."[22] For advocacy groups, the commission's pronounced adoption of the discourse of recovery was viewed as a sign that their long-standing emphasis on the needs and rights of consumers was finally making its way to the upper echelons of public policy. Shortly after the release of the New Freedom report, a broad coalition of nonprofit, consumer-oriented groups organized an "advocacy partnership," reasoning "that if the mental health community could mobilize its advocacy resources around a set of policy objectives designed to move toward the goals outlined in the commission's report," the goals of recovery would be promoted.[23] The partnership came together as the Campaign for Mental Health Reform (CMHR), and includes the National Alliance on Mental Illness, the National Mental Health Association, the Bazelon Center for Mental Health Law, the National Empowerment Center, and the Depression and Bipolar Support Alliance. The CMHR also includes many provider organizations, most notably the American Psychiatric Association. The central goal of the CMHR is to "show the rest of the policy-making world that the mental health field stands behind the Commission's recommendations."[24]

The CMHR illustrates the way in which the discourse of recovery facilitates an alignment between state, professional, and advocacy interests, providing a "common language" for identifying problems and addressing concerns. It is telling that the commission is so often lauded for its rhetorical contribution in disseminating the language of recovery, rather than for any particular policy change. The report includes recommendations for "changing the language connected with mental illness" as a critical way to affect health policy. Michael Stoil documents that after the release of the report, "across the nation, at the federal and state levels, policymakers are adopting the language of recovery and resilience to apply in mental health and mental illness."[25] As discourses of recovery circulate across policy contexts, they situate state and advocacy organizations on "the same side of the fence," enabling a shared language for speaking of mental health care policy.

The New Freedom Commission defines recovery in the report as "the process in which people are able to live, work, learn, and participate fully in their communities," which includes "the ability to live a fulfilling and productive life" as well as "reduction or complete remission of symptoms."[26] This definition is noteworthy for two reasons. First, it draws from a medical lexicon but offers a conception of recovery that is not limited to a narrowly defined "illness versus health" schema. Recovery involves medical processes of symptom reduction, but it reaches beyond this narrow definition of health to incorporate virtually all aspects of life, from occupation and education to the abstractions "fulfillment" and "productivity." Second, the definition emphasizes the necessity of proactive mental health consumers. In order to recover, individuals must be productive and active in the context of their communities. The very definition of recovery, which draws on medico-scientific terminologies, encompasses neoliberal conceptions of individual responsibility. Recovery requires, by definition, a context of care in which individuals are self-directed and in charge of their own choices. Throughout the report, this requirement is simultaneously articulated as a medical necessity grounded in neuroscientific research and as a right heavily imbued with moral overtones. This participation in both scientific and moral registers is what makes recovery a potent discourse of governmental rationality, as exemplified in the National Alliance on Mental Illness's praise of the "new, recovery-focused system": "We believe this effort is justified on moral grounds alone, but is also sound public health policy."[27] The discourse of recovery rationalizes government on moral grounds— recovery-related initiatives are for the good of the people—and on scientific grounds, tapping into values associated with the scientific method, including evidence and objectivity.

Recovery as a Neuroscientific Rhetoric

The New Freedom Commission was launched as an explicitly medico-scientific project. In the executive order creating the initiative, Bush charged the commission with a scientific mission, to consider "how mental health research findings can be used to effectively influence the delivery of services."[28] This research includes a variety of disciplinary knowledge that can claim the status of science, ranging from sociological and psychological studies to economic data. The sciences of the brain play the most prominent role in this "research" assemblage. Bush's 2002 introductory speech paid heed to the social welfare dimensions of the commission, but it drew on medical terminologies by framing mental illness as a "disease" that is treatable, as proved by scientific research that has led to "new drugs and therapies." The makeup of the commission reflected this grounding in medical and brain science, as it included a professor of neuropsychiatry and other research scientists and medical doctors. The neuroscientific contributions to the commission are evident throughout the report, which references research on the brain, neuroscientific advances in treatment, and psychiatric advances in diagnosis and classification, and cites from such journals as *Archives of General Psychiatry*, *Journal of the American Medical Association*, and *Biological Psychiatry*. The language of the report crafts a strong association between brain science research and the rhetoric of recovery, referencing "new understanding of the brain" as a source of information about the "capacity for recovery."[29] The report is peppered with terminologies such as "evidence-based practices" and "science-to-services," all of which solidify the scientific ethos of the commission and its agenda.

The scientific language of the brain is a powerful means of articulating as real both mental illness and recovery. With regard to the reality of mental illness, brain research and especially brain imaging data give mental illness an ontological weight, validating the brain as a space for individual and social interventions. This reality founds claims that mental illness is the same as physical illness, an articulation deployed in political contexts in favor of parity in insurance coverage and reimbursement. The mental illness/physical illness equation also imbues mental health policy initiatives with normative dimensions, making possible the claim that aversion to mental illness initiatives is a form of discrimination or stigma that is scientifically groundless and socially noxious. Thus mental health policies gain scientific and moral validity through neuroscientific proof that mental illnesses are brain diseases. In fact, many advocacy organizations have adopted slogans such as "mental illnesses are brain diseases," or refer to mental illness and brain disorder as synonymous, drawing from neuroscience to articulate moral claims.

In addition to supporting the reality of mental illness, neuroscience is also utilized to frame the recovery perspective as scientifically valid. Brain imaging is also powerful here, because it can depict biological changes purportedly caused by various biological and social recovery–oriented interventions. Neuroscientific research is used to support the efficacy of psychotropic medications, but it is not limited to drug treatments. In the commission's report, the authority of science is used to support a number of interventions as biologically relevant. For example, the commission reports that "emerging research has validated [the proposition] that hope and self-determination are important factors contributing to recovery."[30] Further, the "evidence" shows that "offering a full range of community-based alternatives is more effective than hospitalization and emergency room treatment," because these alternatives encourage consumers "to participate in appropriate and timely interventions."[31] In these examples, the evidence of science is used to frame individual attitudes consonant with a neoliberal framework (hope and self-determination) and individual behavior (participation in accepted interventions) as an essential part of the path to recovery. In a sense, these attitudes and behaviors are elevated to the same status as biological treatments, because all interventions—from attitudinal to pharmaceutical—promote recovery by acting on the brain.

Shortly after the release of *Achieving the Promise*, the Depression and Bipolar Support Alliance (DBSA), one of the advocacy groups mentioned previously who rallied in support of the New Freedom Commission, constructed a follow-up report, *The State of Depression in America*. The paper was intended to augment the commission's recommendations with a renewed emphasis on "the urgent need for action" and continued focus on resolving the barriers presented by fragmentation, stigma, and public ignorance.[32] The DBSA report echoes the language and themes of the New Freedom report, providing a powerful example of how the shared language of mental health informs state and advocacy rhetoric alike. Like many advocacy groups, the DBSA places heavy emphasis on the claim that mental illnesses are "real," biologically based diseases of the brain. Further, the DBSA invests considerable space in supporting the recovery perspective as a scientifically validated mandate for community care programs. The report insists, "Depression is a real illness." Two PET scan images are reproduced, accompanied by the explanation that "the physiological nature of depression is depicted in Exhibit 2, which shows PET (positron emission tomography) scans of a patient's brain during depression and after recovery from depression." After the scans, the report explains, "functional brain imaging studies show alterations in blood flow and metabolism in the brain in some individuals with depression," and the report continues to identify psychosocial and environmental factors that might contribute to the disease in addition to innate genetic factors.[33]

The report depicts two brain scans labeled "depressed brain" and "recovered brain," with the caption, "Image of Normal and Depressed Brain Scan." The two images show marked visual differences in their shading patterns. In their full-color format, the prominently labeled depressed brain is almost entirely dark, with deep blues and purples blending into a black background. There are three or four tiny squares of yellow in isolated fringes of the brain. The recovered brain is generously spattered with yellow, orange, and red regions that suggest the "lighting up" motif associated with brain activity. The side-by-side placement of the two images highlights the contrasts between dark and light, passive and active, monotonous and variegated.

The visual images and their verbal context suggest both the physiological reality of depression as a real illness and the validity of biological treatments. The caption describes the two brains as "depressed" and "normal," illustrating the neuroscientific proof that depression is a real illness that can be seen at the level of brain biology. In the image box, the brains are labeled "Depressed Brain" and "Recovered Brain," providing concrete, visual evidence of recovery. These labels suggest that the images depict the same brain, visualized at different times—the first during the illness, and the second after successful treatment. The use of the singular "patient's brain" in the textual description of the scans bolsters the implication that it is one single person's brain depicted across the time span of his or her illness. The shift from caption to image box from "normal" to "recovered" invites an equivocation between the two terms, such that recovery is synonymous with "normality," and that recovery is synonymous with cure, a medical concept suggesting the elimination of illness through biologically based treatments such as medication.

The DBSA report, like the New Freedom report and similar documents, bolsters this medico-scientific interpretation of recovery by emphasizing scientific development of and consumer access to psychotropic medications. The New Freedom report, for instance, calls for access to "evidence-based, state-of-the-art medications," and the speedy delivery of new drugs to consumers through accelerated "science-to-services."[34] The terms "cure" and "treatment" are often framed as biomedical concepts in a strict sense, implying consumption of pharmaceuticals as the route to restored health. Despite the prominence of scientific and neuropsychiatric references in this discursive assemblage anchored by the New Freedom Commission, recovery is deployed with a significance that goes far beyond strictly biomedical meanings. In the next section, I trace the ways in which the rhetoric of recovery, infused with the authority of visual brain images, facilitates the arrangement of mental health care as a seamless and continuous marketplace (a space dubbed "community") in which consumer-patients act out their own cures through endless processes of choice and consumption that promise to maximize health, productivity, and fulfillment.

Mental Health Care, the Personal Journey of Recovery, and Neoliberal Economics

The rhetoric of recovery frames reductions in state services as moral victories. In this articulation, the state is not abandoning individuals to scramble for their own supports, but rather is empowering them with the means of recovery—independence, responsibility, and choice. Recovery, in these formulations, is not simply a scientifically objective state of health, but also a deeply personal quest driven by individuals' most authentic desires and unique capacities for fulfillment and productivity. Definitions of recovery reflect this personalization. For example, in 2004, SAMHSA hosted a conference devoted to coming up with a standard definition of recovery. The National Consensus Conference on Mental Health Recovery and Systems Transformation featured more than 110 experts, including mental health consumers, providers, and government officials. The conference determined the following definition of recovery: "Mental health recovery is a journey of healing and transformation enabling a person with a mental health problem to live a meaningful life in a community of his or her choice while striving to achieve his or her full potential."[35] The metaphor of a journey frames recovery as a narrative that organizes each individual's life into a unified sequence, assimilating past, present, and future into the mental health quest. In the journey metaphor, mental illness is not something that intrudes on and ruptures an individual's life from the outside, but rather a formative development internal to each person's life that draws together and gives coherent shape to the sequence of their story. Other definitions of recovery resonate with the journey metaphor, framing mental health as a goal that is both comprehensive, in that it permeates and draws together all aspects of life, and personal. For example, a recent review of definitions of recovery published in the *Psychiatric Rehabilitation Journal* concludes that recovery is "multidimensional, fluid, non-sequential, complex, and permeates the life context of the individual."[36] What definitions of recovery have in common, then, is the sense that mental health is a subjective and personal determination, as well as something tied to scientific and medical objectivity.

If what constitutes mental health is a personal, subjective determination, it follows that mental health care must be responsive to the diversity of conceptions of mental health recovery that emerge from individuals' unique journeys. In the recovery movement, mental health care itself is premised on the themes of personal choice and responsibility, as reforms seek to put consumers in charge of determining what "health care" means to them. The New Freedom report defines mental health care as an extraordinarily broad domain: "The mental health care system collectively refers to the full array of programs for anyone with mental illness. These programs exist at every level of government and throughout the private sector. They have varying missions, settings, and financing. They deliver or pay for treatments, services, or other types of supports, such as housing, employment, or disability benefits. . . . The setting could be a hospital, a community clinic, a private office, a school, or a business."[37] Mental health care, in other words, is not unified by virtue of its spatial location or even the type of treatment it offers—therapies ranging from buying shoes to occupational coaching to medications are all potentially a part of mental health care, whether they occur at home, at school, or in a more traditional medical setting. What defines mental health care and unifies it into a coherent system is the individual consumer. Whatever treatment an individual selects, in whatever social space the individual desires to travel, all are linked together as a mental health system solely on the basis of the individual's personal choice. The individual consumer is conceived as the center of a radiating network of services in a "seamless" system that overcomes institutional fragmentation.

The rhetoric of recovery's elevation of choice and individuality is made concrete in federal programs that encourage states to adopt mental health care initiatives that put individuals in charge of creating their own programs. The New Freedom report emphasizes that financing schemes must be highly flexible to facilitate recovery. Flexibility of financing means that individuals are responsible for allocating allotted funding to purchase the services of their choice, through the aid of "brokers," and to construct "a personalized, highly individualized health management program." Consumers must "play a larger role in managing the funding for their services and supports." With transformation that places funding under the management of individual consumers, "incentives will shift toward a system of learning, self-monitoring, and responsibility," and individual consumers will have a "vested economic interest in using resources wisely to obtain and sustain recovery."[38] The vision of the New Freedom Commission is promoted by federal grant programs, including Real Choice System Change Grants and Mental Health Transformation State Incentive Grants, which create incentives for state and local programs that put consumers in charge of selecting and coordinating their own mental health care services.

Florida SDC is just one example of a program that operationalizes these imperatives to maximize choice and economic efficiency. For example, another program praised by the federal government is Wraparound Milwaukee. Like Florida SDC, Wraparound Milwaukee implements a flexible, consumer-centered funding model. A diverse array of services is coordinated by a central agency, and funds previously allocated for institutional care are combined in a single pool and used to fund community-based services directed by and personalized for each individual. The consumer-centered program is heralded because it is scientifically validated (or "evidence-based"), socially productive, and cost-effective. "Funding streams" from different sources are pooled together and then placed in the hands of individuals, who craft personalized recovery programs by selecting and purchasing services that "wrap around" the entirety of their unique lives, taking into account identity factors such as race, gender, and cultural background.

Although consumer-centered programs are praised because they value diversity and facilitate empowerment and choice, they also issue an underlying moral judgment tied to the social value placed on economic productivity. In the recovery discourses, a healthy, recovering individual is above all economically productive. Thomas Nerney of the Center for Self-Determination describes how recovery-oriented systems address an underlying root of mental health problems: the lack of personal motivation exemplified in the question "Why get up in the morning at all?" Nerney writes that recovery's emphasis on self-determination works to assist individuals in creating meaningful lives "deeply connected to their communities and the world of business and commerce." Through recovery-oriented programs, individuals are encouraged to "carry out daily responsibilities, work, earn income, plan for life goals, take care of family members, contribute to the common good, and exercise citizenship."[39] Individuals who are unable or unwilling to work at "typical jobs" can carry out recovery by developing "microenterprises," or alternative means of productively contributing to the business and commerce of their communities.[40] Thus a fundamental component of the recovery paradigm is an injunction to contribute to society in a way that is measurable in economic terms. Individuals who fail to contribute economically are guilty of being unhealthy. As the term guilt implies, this failure is coded as both pathological and morally condemnable, a refusal to take up one's duty to be well. As in other brain culture discourses, this failure is situated in relation to both society and one's own self—a failure to recover is to refuse to maximize one's contribution to society and realize one's most authentic personal fulfillment.

Consumer-oriented mental health care programs position individuals as active participants in the marketplace, and they facilitate these patients' assimilation into the commerce of their community by preparing them for active roles as producers, or entrepreneurs of themselves who maximize at once their own potential and the economic productivity of society. This preparation includes instruction in risk management. Self-directed care programs carry considerable risk—individual consumers are allocated limited resources, and if they choose the wrong services out of ignorance, ill planning, or sheer capriciousness, they may be unable to acquire the services they truly need. In the recovery discourses, these risks are framed as a positive value because they promote genuine economic participation. A report by the UPenn Collaborative on Community Integration, funded by the National Institute on Disability and Rehabilitation Research, states, "SDC is a calculated risk, and one that we should not be afraid of. True integration means real life risks as well as rewards!"[41] Individuals will learn from their mistakes and become capable of managing risks in the contexts of health and economics. Thus part of the goal of SDC programs is to train individuals to accept responsibility for managing risks to their physical, mental, and economic well-being, and to govern their own lives in ways that optimize their social and economic productivity. By implication, if a consumer is unable to piece together a mental health care program that maximizes their social contribution and personal fulfillment, they are the one responsible.

By training individuals in risk management and self-optimization, the SDC programs valorize market-oriented programs as scientifically valid, economically efficient, and morally justified. For instance, SAMHSA's *Free to Choose* report states, "A rational service delivery system is one that delivers care that is timely, effective, and efficient—that is, care that achieves what it intends with the use of the fewest resources."[42] This is only possible when individuals are in charge of their own care, because only then will they treat funding "as an asset rather than an entitlement." Implicit in SAMHSA's description is the idea that social services are tied to moral corruption—they are taken as "entitlements" and essentially let individuals off the hook for not contributing productively to society. This moral corruption can be corrected when individuals are literally forced to be responsible for their own care, to recognize the economic value of the services they receive, and to pay for these services with their own economic contributions.

Conclusion: The Rhetoric of Recovery and Entrepreneurial Selves

Policy reforms tied to the recovery movement prioritize the maximization of human capital, conceived in various healthist terminologies. Mental health care can be understood in Foucault's terms "as so many elements which enable us, first, to improve human capital, and second, to preserve and employ it for as long as possible."[43] Individuals are encouraged to live their lives according to a logic of enterprise, where even the consumption of goods and services is a form of production aimed at maximizing personal satisfaction. That satisfaction is accorded economic value, as it is tied in various ways to social productivity and good citizenship. As the rhetoric of recovery illustrates, entrepreneurial discourses are most effective when they speak in languages that are not strictly economic or scientific: recovery is powerful because it speaks in terms of community, democracy, and personalization in order to, in Foucault's terms, compensate for "what is cold, impassive, calculating, rational, and mechanical in the strictly economic game of competition."[44]

The rhetoric of recovery is a successful compensatory discourse, in the sense Foucault describes: it is an accommodating and an enticing rhetoric that can promise, along with medical cure and scientific authority, the democratic ideals of participation and access, freedom and choice, antidiscrimination and respect for individual diversity. The problem with this warm-hearted rhetoric is that whatever social ideal or personal desire it articulates, it ends up validating the free market and, more specifically, the enterprise model as the obvious solution to whatever ails, whether the ailment is conceived in medical, moral, or economic languages. As Jodi Dean argues, rhetorics of access, participation, and democracy can "work ideologically to secure the technological infrastructure of neoliberalism."[45] Moreover, when everyone comes to speak the same languages of freedom, equality, and participation, then we come to believe that there is no alternative to neoliberalism and its market-oriented solutions. It is difficult to articulate, Dean explains, a language alternative to one of democracy and participation, one that has not become thoroughly soldered to the ideals and principles of the market.

My point in this chapter is not to argue that we should go back to a different model of social services, where the state happily provides mental health care services to patients who are not expected to become producers of their own satisfaction. Nor is my purpose simply to indict emerging programs of care oriented toward market-based and entrepreneurial models. A consideration of whether these programs benefit individuals is beyond the scope of this chapter. My purpose is, rather, to draw attention to the extent to which mental health care is bound up with economic modes of thinking, and to trace out the ways in which mental health rhetorics tie together moral dimensions of health, economic efficiency, and scientific authority. Specifically, I highlight the ways in which mental health care has become a site for training and motivating self-entrepreneurs, to the point where mental health itself is defined in terms of market-based activities. In the next chapter, I conclude my inquiry into brain culture by developing this line of thinking, exploring what is at stake in popular neuroscience and its injunctions to optimize one's self in the name of personal freedom, economic productivity, and scientific truth.

7

The Brain Is the Frontier

The Subjectifications of Neuroscience

As evidenced by its pervasive circulation across a variety of cultural forms, the language of the brain is a powerful vocabulary through which citizens today express their anxieties, articulate their hopes and dreams, and rationalize their disappointments. This neuroscientific grammar has not replaced other languages, such as those of religion or psychology, but as its force in public life accelerates it continues to infiltrate other strains of discourse, casting neurological significance on spiritual, familial, educational, economic, and other social idioms. These brain-based ways of understanding self and society disseminated by popular neuroscience are entangled within a broader narrative that situates scientific knowledge as a means of achieving mastery over human nature, and ultimately of realizing utopian visions of total personal fulfillment and universal social harmony. In this concluding chapter, I explore this narrative of scientific mastery and offer an alternative lens for conceptualizing popular neuroscience and brain culture as rhetorical phenomena. Instead of understanding brain science as a set of facts that must be true or false, I view scientific conceptions of the brain as a rhetorical space of sorts, where diverse social actors work through fundamental questions about what it means to be human in particular historical and cultural contexts. Rather than judging the truth or falsity of this discursive activity that centers on the brain, I attend to its cultural effects. Specifically, I conclude by revisiting a theme that runs throughout the case studies of preceding chapters: that the discursive activity of popular neuroscience works through questions of human nature in ways that facilitate neoliberal social and economic arrangements, and that contribute to the guilt and anxieties associated with relentless imperatives to optimize self and society.

The Brain Is the Frontier

Popular neuroscience's narrative of scientific mastery is often articulated through the trope of the frontier. In the most common deployment of this trope, the brain itself is figured as a "frontier," or a "great uncharted territory" that has yet to be mapped, the "last and greatest" biological frontier that is like a planet awaiting exploration. In relation to this neural "frontier," brain scientists are depicted as the explorers, the great Magellans who will map the brain through imaging technologies and conquer the frontier by rendering it transparent to scientific knowledge. At the end of this conquest, the brain explorers will know "all there is to know about human nature and experience."[1] James Watson describes scientists as "discoverers" who have found the brain through their explorations: "The brain is the last and greatest biological frontier, the most complex thing we have yet discovered in our universe."[2] In these examples, scientific knowledge of the brain is framed as complete and total knowledge of human being. The brain, according to Watson, is the "last" frontier; upon its conquest, nothing will remain unknown or hidden from the eyes of science.

In his address at the baby-brain summit discussed in chapter 4, President Bill Clinton gives an explicitly American spin to this familiar trope. Brain imaging research, Clinton states, "has opened a new frontier. Great exploration is, of course, not new to this country. We have gone across the land, we have gone across the globe, we have gone into the skies, and now we are going deep into ourselves and into our children."[3] In his speech, Clinton frames brain imaging research as something uniquely American and somehow integral to our identity as a nation. He subtly draws on cultural knowledge about American history, referencing Manifest Destiny ("across the land"), the country's founding by courageous explorers ("across the globe"), and another aspect of history that has drawn considerably from the language of the frontier: space exploration ("into the skies").

Not only does Clinton's ode to brain science frame scientific endeavors in terms resonant with all the mythos of American history, it ties the brain to human subjectivity and frames both in spatial terms as territory to be conquered. Visualization of the brain through imaging technologies is equated with introspection and understanding of human personality. Going "deep into" our selves and our children implies that the core truths of our being are revealed with digital brain images. By identifying scan with psyche and situating this visualization with other great moments of exploration and conquest, Clinton's use of the frontier trope flattens out the interior of human subjectivities and constructs our "selves" as accessible spaces. In other words, the human psyche is not something hidden deep within, an intangible and intimate interior, but something that can be seen, known, and manipulated. Perhaps most important, it can be explored topographically. In short, in Clinton's narrative there is a rhetorical movement that transforms the metaphoric space of psychic interiority, or subjectivity, into the literal space of the physiological brain.

The brain as frontier narrative equates the space of subjectivity with the space of the brain, and insists that scientific knowledge will bring about total control—and, through this control, ultimate fulfillment and peace. This narrative achieves bold formulation in Zack Lynch's *The Neuro Revolution*, where he anticipates "a radical reshaping of our lives, families, societies, cultures, governments, economies, art, leisure, religion—absolutely everything that's pivotal to humankind's existence." This "gigantic wave of transformation," Lynch continues, will "reach every corner of the planet." If humans keep their aspirations high and approach this wave with "benevolent intelligence," the results will be extraordinary: "It will let us create a future of greatly enhanced, better-balanced, and more satisfying individual lives within a vastly transformed society that we will build through an unimaginably powerful capability we've never had before: increasingly precise control over the most complex entity in the universe, the single most important determinant of the quality of the lives we lead—our human minds."[4] Knowledge of the human brain will guarantee "more solid and reliable decisions" and "more lasting happiness." Humanity will come to live "comfortably, harmoniously, and prosperously," as they come to control both the worlds they inhabit and the worlds within themselves. Lynch's utopian vision is echoed by Richard Restak, who sees neuroscientific knowledge as a way to "advance human freedom" and move down the "path toward the achievement of both personal and collective liberation."[5] Similarly, in *Brave New Brain*, Nancy Andreasen also sees neuroscience as the source of "enlightenment and knowledge" that can "conquer" pain and suffering, and build "healthier, better, braver brains and minds."[6]

What these articulations have in common is two related assumptions. First, they assume that knowledge inevitably leads to control, and specifically that scientific knowledge will lead to greater capacities to intervene in human nature and design it through planned brain-building initiatives. Second, these discourses assume that human control inevitably leads to positive outcomes. If humans acquire the capacity to control and manipulate their own nature (including their identities, societies, and worlds), then they will bring about personal happiness and social peace, as well as liberty, freedom, and universal harmony. Together, these assumptions constitute a teleological narrative that views humans as progressively increasing their knowledge through scientific discovery, moving closer and closer to total mastery, which is taken as equivalent to total liberation. Humans are, in this narrative, fundamentally good, and the only barrier to the realization of personal and collective harmony is a lack of knowledge. Brain science, in this story, is accorded a quasi-divine power, as its capacity to reveal the truths of human nature is the key to a type of secular salvation that can be achieved by scientific knowledge as it makes the ultimate reaches of human being transparent to itself. What is interesting about this underlying teleological narrative is that it cannot be validated by scientific fact—it is, rather, a broader rhetorical framework through which scientific research is interpreted and made sense of in relation to culture. Thus popular neuroscience's affinity for utopian visions shows the extent to which investigations into the brain are about more than science or fact—they are discursive activities through which cultures work out their conceptions of their selves, their worlds, and their futures.

This narrative is no doubt a hopeful one, because it anticipates an end to all human suffering, both physical and psychological, from personal distress to the collective agonies that result from global violence and poverty. At the same time, however, this narrative has the capacity to generate immense guilt and anxiety because, when translated into practical guidelines for daily living, it issues relentless imperatives to self-optimize. The utopian tale works on two levels: Science will acquire total knowledge of human nature through its inquiries into the brain, and this knowledge will enable the construction of what Lynch describes as "nothing less than the birth of a new civilization."[7] Achievement of this new civilization requires, however, activity at a second level—work by individuals at the level of everyday life. According to popular neuroscience, activity at this level must occur now, in preparation for and as a means toward greater societal transformation. Even as scientists push forward in their quest for total mastery of the brain and mind, brain research already knows enough to provide individuals with the knowledge necessary to achieve their own highest levels of happiness and fulfillment. For example, Rick Hanson encourages individuals to take advantage of scientific wisdom and "reach inside" their own brains "to create more happiness, love, and wisdom."[8] The result is not only kinder, happier, more successful individuals, but also a better world. Hanson continues, "As you and other people become increasingly skillful with the mind—and thus the brain—that could tip our world in a better direction."[9] The great social transformation that Restak and Lynch herald requires the concerted action of individuals who faithfully come to know and act on their own brains to produce themselves as better citizens.

As previous chapters have described, good citizenship is often defined in terminologies of health, as various positive life characteristics—success, happiness, and relational satisfaction, for instance—are assimilated into definitions of "healthy brains." Moreover, because popular neuroscience promises individuals that suffering, conceived in healthist vocabularies as different varieties of illness or neurological deficiency, can be ameliorated, it situates individuals as responsible for their own failures to be free, happy, and fulfilled. Freedom and fulfillment are not simply promises; they are imperatives, duties to both society and, more intimately, to one's self. Failure to meet these duties occurs in relation not only to one's social world, but also to one's most authentic self and truest desires. Because a failure to be well—whether conceptualized in biological, social, or personal terms—is framed as betrayal of one's own self, it is capable of generating a unique intensity of guilt and anxiety.

As the case studies trace out, popular neuroscience works to both generate this anxiety and channel it into habits of living focused around maximizing brain health. Individuals are encouraged to constantly attend to the brain: "What is the state of my brain? What are its shortcomings, and what are its potentials? How can it be improved, and what puts it at risk? How can I make myself healthier?" Because of the ways in which popular neuroscience configures the brain (and the self) as a plastic entity amenable to influences of all types, it turns almost every aspect of living into something that must be accounted for in terms of its influence on both the brain and the self, its contentment and productivity. To borrow the words of Peter Miller and Nikolas Rose, popular neuroscience multiplies "the points at which normative calculation and intervention are required."[10] There is nothing that cannot be viewed in terms of its relation to the brain, whether that relation takes the form of "How will this affect my brain?" or "What does this effect tell me about the way my brain is working?"

When the metaphorical space of subjectivity is equated with the material space of the brain, there are no reaches of human identity that are exempt from logics of calculation and intervention. For example, religion is commonly understood as an aspect of one's soul, the most sacred and authentic of interiors. In the context of neuroscience, however, religion is articulated as a product of neurological activities that can be manipulated in order to produce more healthy religious experiences. For instance, Michael Persinger has gained attention in popular media for designing a helmet that emits electromagnetic pulses to the brains of those who wear it, hypothetically inducing religious and spiritual experiences. Persinger's studies complement those of Andrew Newberg, whose use of brain imaging to locate the "seat" of spiritual and religious experiences has gained considerable notoriety in both religious and secular circles.[11] And in a 2004 *Time* article, Jeffrey Kluger reports on studies that link religious beliefs to the very same brain chemicals that are manipulated by Prozac and other antidepressants.[12] In these discourses, what many might cling to as the utmost core of their being—the essence of their interiority—is articulated as something that can be disassembled or reassembled through electric and magnetic stimulations.

For the most part, these studies of the neural basis of religious experience do not insist that religion is inauthentic or invalid. The focus, rather, is on the ways that religious experience can be induced, manipulated, and channeled to foster greater health, productivity, and happiness. One example is Daniel Amen's work, discussed in chapter 3, which views brain health as the basis for positive religious experiences. While Amen tends to view religion through the lens of the Christian tradition, Hanson's book *Buddha's Brain* recommends brain-based habits of living to achieve spiritual peace and enlightenment, and Sharon Begley's book discussed in chapter 2 also makes much of the connections between practical neuroscience and Buddhism. In all these books, the theme is that religious health can be restored and even amplified through actions on the brain. In addition, religious satisfaction is valuable because it is a key means of maximizing health and well-being—religious experience does not necessarily have intrinsic value, but it is important because it helps individuals to optimize their life potential. For example, in their book *How God Changes Your Brain*, Andrew Newberg and Mark Waldman argue that religious experience can reduce stress, promote well-being, and slow down the aging process, conserving the resources of health. Regardless of whether they are true in any ultimate sense, spiritual practices are valuable because they "enhance the neural functioning of the brain in ways that improve physical and emotional health." These practices strengthen a neurological circuit "that generates peacefulness, social awareness, and compassion for others."[13] Thus religion is valuable because of its practical significance in terms of accumulating greater health and contributing to social harmony. Individuals who bolster their health through spiritual practice will be more socially aware and compassionate because they exhibit the neural changes "we need to make if we want to solve the conflicts that currently afflict our world."[14] Individuals who engage in religion can be assured that even if their views are ultimately mistaken, they still "bring a little more peace into the world."[15]

As the example of religious experience suggests, there is no aspect of life that is natural, in the sense of taken for granted, predetermined, or fixed beyond individual and social control. Religion is put in the hands of individuals—rather than a divine gift, or a reality disclosed by nature, it is a practice, or set of practices, that individuals can choose to engage in so that both they and their society can reap the benefits of religious experiences and associated qualities of peace, patience, and kindness. In this articulation, humans are conceived as the authors of their own fate in every matter, whether that be job success, family stability, sexual pleasure, or religious satisfaction, and in all cases, fate must be secured by way of the brain. All these areas of life become

places that demand focused efforts—scrutiny, calculation, and intervention—
in order to maximize health reserves and stave off health risks. Whatever
the area of life, failures cannot be chalked up to some external fate, whether
natural or theological—individuals are, popular neuroscience drives home,
responsible for cultivating and optimizing their own brain health. Moreover,
all these different areas of life acquire their value in relation to the health of
the brain—even something as hallowed as religion is assessed in terms of its
abilities to increase individual and social health.

The critical point about popular neuroscience is not simply that it generates anxieties and a sense of guilt through its "relentless imperative of risk
management," applied to "choices of where to live and shop, what to eat and
drink, stress management, exercise, and so forth."[16] These affects—anxiety
and guilt, as well as hope and anticipation—have an economic value in the
sense that they play a vital role in neoliberal economics. These affects turn
individuals into entrepreneurs of themselves, agents who calculate the consequences their investments will have on their happiness, productivity, and
fulfillment and intervene accordingly. Because these entrepreneurial models
are articulated in neuroscientific languages, they acquire a veneer of truthfulness and obviousness that is not provided by solely economic discourses.
This is perhaps the most important reason to approach popular neuroscience with a critical eye: whatever benefits it might provide to individuals
seeking to improve their lives, it naturalizes ways of thinking and corollary
social and economic arrangements that can and do produce deeply trouble-
some effects. A thorough critique of neoliberalism is beyond the scope of
this book, but previous chapters have sketched out the extent to which
popular neuroscience resonates with, and in many cases directly feeds, these
arrangements. While many others have focused on the economic, environmental, and social costs of the dominance of market-based models, in this
book I am more attentive to the costs that are felt in everyday, mundane
existence as individuals feel obligated to the ceaseless pursuit of ever-
greater quantities of happiness and health.

The critical studies of this book have tended to focus on the ways in
which popular neuroscience ties individual lives to entrepreneurial models
of living. For me, this is one of the most significant issues related to popular
neuroscience's articulation of the brain. In academic discourse especially,
critiques of biological determinism have become commonplace. To simplify
matters a great deal, these critiques target determinist theories for natural-
izing problematic social and political arrangements. Moreover, conventional
criticisms charge that these determinist theories tend to stymie conceptions
of social change because they associate individual and cultural events with
external forces that are, for the most part, impervious to human will and
intention. But in the case of popular neuroscience, even though biological

languages are prominent, human will is virtually unfettered in its ability to change the brain and bring about far-reaching personal and social transformations. Popular neuroscience shows, then, that theories of human will and social agency carry their own risks. These theories might break the chains of biological determinism, but in its place they issue a relentless command for personal responsibility accompanied by unremitting obligations to accumulate more health, achieve more productivity, and enjoy more happiness. These new chains, I suggest, are as heavy as the chains of determinism, although they work very differently. I am not arguing we should flee back to the relative security of determinist conceptions of human life, but rather that it is important to recognize that both determinist and agency-centered paradigms bring with them their own respective benefits and dangers.

Turning the Trope: The Brain Is the Frontier, Reconsidered

Utopian visions encompassing dreams of personal fulfillment, economic productivity, and social harmony are communicated through discourses that equate brain and self, and view both as a frontier or territory to be mastered and tamed through the inevitable progress of human knowledge. This vision depends on the assumption that scientific knowledge, especially that acquired by imaging technologies, makes the truth of human nature transparent and hence controllable. In the remainder of this chapter, I want to take the popular metaphor of the brain as frontier and turn it, or read it in the context of a different framework. In popular neuroscience, the brain-as-frontier metaphor implies a colonialist understanding—as an object of scientific knowledge, the brain (and hence the entire person) is a passive territory, and scientists (as well as citizens who take up scientific knowledge) are conquerors who civilize this frontier through their willful efforts. Instead of viewing the brain as an object that is increasingly revealed to science, I understand the brain as the focus of discursive activity that functions to work out significant questions about identity and social relations. Throughout history, investigations of human nature, social order, and identity have taken place through discussions and theories about the brain. The various ways the brain has been articulated in distinctive historical and cultural contexts resonate with social beliefs about the nature of human being and prevailing modes of social, political, and economic order. Thus, from a rhetorical perspective, theories of brain function are interesting not primarily because of the truthful facts they reveal about the world, but rather because they function as a key locus for cultural discourse about self and society. As a space of discursive activity, the brain can be read as a different sort of frontier—not a territory to be conquered, but a mobile and dynamic space that changes its colonizers as much as it is affected by their quest for ultimate knowledge and control. I draw from Frederick Jackson Turner's quintessential discussion of the frontier to characterize the brain as a dynamic rhetorical space. Turner's theorizations of the frontier have sparked numerous responses and diverse inquiries; here, my purpose is to briefly engage Turner's scholarship to suggest how the frontier can be conceptualized as a zone of rhetorical activity that resists ultimate conquest and total knowledge.

Turner's contribution to American history is his attention to the uniquely American construction of the frontier. In Europe, the concept of frontier designates a determinate and stable political boundary, but in the United States, with its vast geography and particular legacy of westward expansion, the frontier signifies the edge of a settlement, a mobile line that constantly reconfigures the elements on each of its sides. Turner writes, "Thus American development has exhibited not merely advance along a single line, but a return to primitive conditions on a continually advancing frontier line, and a new development for that area. American social development has been continually beginning over again on the frontier. This perennial rebirth, this fluidity of American life, this expansion westward with its new opportunities, its continuous touch with the simplicity of primitive society, furnish the forces dominating the American society."[17] Turner's description is romantic to be sure, but it gets at several important functions of the frontier that are often neglected in contemporary critiques of colonial discourse. First, the frontier is mobile—a line or space that is continually advancing. Unlike the brain scientist's confidence that investigations into the brain will ultimately lead to the end of unknown nature, in the sense that everything will be known and revealed once and for all, Turner's space is permanently capable of new movements.

Second, the movement that Turner describes is not simply an assimilation of more territory to the colonizer's (or pioneer's) way of life. The conditions of the latter are fundamentally transformed by the former—the way of life, or the identity, of the pioneer is constantly being reinvented and reconfigured. The pioneer's "subjectivity" is fluid and is transformed as much as the geography of the territory is by the continuous dynamism of the frontier line. It is impossible to conceive of the movement of the frontier as a unilateral exercise of power or a relation of mastery. At the frontier, the pioneer "must accept the conditions which it furnishes, or perish."[18] Power is reciprocally conditioned by the frontier space itself; instead of an interpretation that attributes sole causality to the pioneer, the frontier itself can be thought of as possessing an agency or force that transforms and articulates both territory and pioneer.

Turner's construction of the frontier lends itself to this conception of an active frontier. For example, he describes the frontier as a "moment" when "the bonds of custom are broken and unrestraint is triumphant."[19] In this passage, the frontier is not a parcel of space but an interval of time marked by its suspension of the established habits of the pioneer. In this moment, the characteristics of the territory seem to overtake the pioneer, a movement of mastery that is reversed from the conventional understanding. This is a definitive moment of transformation, when the habits and customs that have defined the pioneer and his or her "civilization" are thrown into question and the pioneer becomes, as much as the territory, an assemblage of elements awaiting a new articulation.

Turner uses the active voice to describe the frontier's address to potential explorers. It issues an "imperious summons," presenting itself as a "gate of escape from the bondage of the past," despite the fact that it never completely abandons its proximity to custom and inherited ways of doing things.[20] The frontier is the active party, compelling the pioneer to mold him- or herself to its demands. The pioneer who stands at the space of this "gate" is placed in that interval of undecidability, momentarily in between tradition, custom, civilization, and nature. The frontier is this moment of indetermination between nature and culture, and it features a strange and paradoxical reciprocity. At the moment of the frontier, the pioneer is thrown back to a more primitive "natural" state, at the same time that he or she moves forward to impose a new social order on the frontier. In this moment between future and past, in between colonization and past determination, the determination of the individual and of society is thrown into relief. The frontier is an edge that "calls out militant qualities and reveals the imprint of wilderness conditions upon the psychology and morals as well as the institutions of the people."[21]

The frontier mythology is part of the tradition of American individualism, the narrative of the solitary individual battling the forces of nature to acquire freedom and mastery over the environment. But the frontier also functions socially and can be thought of as an agent of dispersion. The frontier, Turner notes, was treated in census reports as the margin of that settlement with a density of two or more per square mile. Turner develops this, describing the frontier as an elastic line partially defined by networks such as communication systems, railroads, populations, and natural markers such as rivers or mountain ranges. Depending on one's orientation, the frontier might differ—he speaks of a farmer's frontier, trader's frontier, and merchant's frontier. The frontier itself is defined in terms of both nature and culture. Turner writes, "Thus civilization in America has followed the arteries made by geology, pouring an ever richer tide through them, until at last the slender paths of aboriginal intercourse have been broadened and interwoven into the complex mazes of modern commercial lines; the wilderness has been interpenetrated by lines of civilization growing ever more numerous. It is like the steady growth of a complex nervous system for the originally simple, inert continent."[22] The frontier is at once a natural space—for example, the natural river that preexists its social use—and a space with social import that is used to transport goods for economic purposes. As the spaces of nature and the spaces of social development become intertwined, the difference between them is undecidable. The river is just as natural and just as socially constructed as the shipping route.

When Turner states that the frontier brings civilization into contact with savagery, throwing the pioneer back to a "primitive" state and calling forth the continual renewal of social systems, he alludes to the indeterminacy of both nature and culture. The pioneer ideal is one of conquest, to be sure: Turner describes a universal quest after the unknown, a yearning to go, quoting Kipling, "beyond the sky-line where the strange roads go down."[23] Yet, as the pioneer pursues this line, he or she is forced to make old tools fit new purposes, to reshape former habits, institutions, and ideas to changed conditions, and to craft new social arrangements as well as break new soil. The frontier mentality, as Turner describes it, entails a rebellious, anti-conventional spirit, but it is not a pure creative power of actively shaping a passive nature. The frontier signifies a grappling, a struggling, a mutual transformation of nature and culture, pioneer and territory, through the movement of a line that constitutes the boundaries of both. The frontier is an individuating edge that defines the contours of territory and pioneer but is itself undecidable.

When the brain is conceived as frontier, it is not simply an untamed biological organ or a pure natural space awaiting scientific discovery. Scientific endeavors to "map" this brain space are constitutive operations that simultaneously transform the "subject" and "object" of knowledge. Gilles Deleuze describes the "entire biopsychic life" in spatial terms, as "a question of dimensions, projections, axes, rotations, and foldings."[24] The brain, he writes, "is not only a corporeal organ but also the inductor of another invisible, incorporeal, and metaphysical surface on which all events are inscribed and symbolized."[25] This is clearly seen in the central paradox of popular neuroscience rhetoric: the brain is both the biological organ that exhibits arrays of electrical and chemical activity, but it is also the surface on which all aspects of subjective existence are inscribed. In other words, the brain is the rhetorical space for discursive activity that works to define human identity and social existence in their most fundamental aspects. In contemporary discourses of the brain, visualizing the electrical and chemical activity of the brain *is* seeing the roots of every whim and behavior. This "bio-metaphysical" brain becomes the discursive site for the articulation of numerous individual and collective social practices (child development, education, self-improvement, health care, etc.). The brain is the object of knowledge, but it is also what drives the quest for knowledge: it is the ulti- mate "knower." This strange configuration of the brain as an entity with dual citizenship in the realm of the biological and the metaphysical leads to such odd statements as this from E. O. Wilson: "The human brain is the most complex object known in the universe—known, that is, to itself."[26] When new knowledge about the brain is produced, the rhetorical brain also changes those who know it, the scientists and others who seek to know the workings of the brain and, ultimately, their own nature.

The rhetorical perspective understanding the brain as a frontier, a discursive space of articulation, is not necessarily mutually exclusive with scientific understandings of the brain. In other words, the rhetorical para- digm does not demand that one give up all claims to scientific validity or the idea that scientific understandings of the brain can and do produce benefi- cial outcomes for individuals and societies. What I find challenging about popular neuroscience is not so much its particular claims to knowledge—for instance, claims that in certain cases psychotropic drugs can reduce suffer- ing, or that focused meditation and education can improve life performance. Rather, what I find problematic is the underlying narrative that humans can achieve total happiness through their scientifically validated efforts, and the entrepreneurial ways of living this narrative supports. It is this narrative that gives the obligations imposed by popular neuroscience such weight—if humans can will their own faultless improvement, then they must bear the crushing responsibility both for achieving this state and for their failures to realize and accomplish personal and social perfection.

NOTES

CHAPTER 1   THE RHETORICAL BRAIN

1. Georgia Andianopoulos, *Retrain Your Brain, Reshape Your Body: The Breakthrough Brain-Changing Weight-Loss Program* (New York: McGraw-Hill, 2008); John Arden, *Rewire Your Brain: Think Your Way to a Better Life* (Hoboken, NJ: Wiley, 2010); Ryuta Kawashima, *Train Your Brain More: Better Brainpower, Better Memory, Better Creativity* (New York: Penguin, 2008); and Madeleine Van Hecke et al., *The Brain Advantage: Become a More Effective Business Leader Using the Latest Brain Research* (Amherst, NY: Prometheus Books, 2010).

2. Nikolas Rose, "Neurochemical Selves," *Society* 41, no. 1 (2003): 46–59.

3. Kawashima, *Train Your Brain More*, 5.

4. Quoted in Tracy McVeigh, "Computer Games Stunt Teen Brains," *The Observer*, August 19, 2001, http://www.guardian.co.uk/world/2001/aug/19/games.schools (accessed April 4, 2010).

5. See Deborah Cameron, "Moving to the Dark Side of the Screen," *Sydney Morning Herald*, May 13, 2006, http://www.smh.com.au/news/technology/dark-side-of-the-screen/2006/05/12/1146940739294.html (accessed April 5, 2010); and Neal Feigenson, "Brain Imaging and Courtroom Evidence: On the Admissibility and Persuasiveness of fMRI," *International Journal of Law in Context* 2, no. 3 (2006): 233–255.

6. For more on both the operations of functional imaging technologies and their persuasive force in public contexts, see Anne Beaulieu, "The Brain at the End of the Rainbow: The Promises of Brain Scans in the Research Field and in the Media," in *Wild Science: Reading Feminism, Medicine, and the Media*, ed. Janine Marchessault and Kim Sawchuck (New York: Routledge, 2000), 39–52; Joseph Dumit, *Picturing Personhood: Brain Scans and Biomedical Identity* (Princeton, NJ: Princeton University Press, 2004); Michelle Gibbons, "Seeing the Brain in the Matter: Functional Brain Imaging as Framed Visual Argument," *Argumentation and Advocacy* 43, nos. 3/4 (2007): 175–188; and Kelly Joyce, *Magnetic Appeal: MRI and the Myth of Transparency* (Ithaca, NY: Cornell University Press, 2008).

7. Steven Johnson, *Mind Wide Open: Your Brain and the Neuroscience of Everyday Life* (New York: Scribner, 2004), 179, 164.

8. Bruno Latour, "Drawing Things Together," in *Representation in Scientific Practice*, ed. Michael Lynch and Stephen Woolgar (Cambridge, MA: MIT Press, 1990), 19–68.

9. AFP, "'Brain Training' Dr. Kawashima Has No Time for Games," January 30, 2008, http://afp.google.com/article/ALeqM5gMwDeIovbiILhtf3JKM2Ez79rGvA (accessed April 5, 2010); Grace Wong, "Boom Times for Brain Training Games," *CNN Health*,

December 11, 2008, http://www.cnn.com/2008/HEALTH/12/11/brain.training/ (accessed April 5, 2010); and Kris Graft, "Nintendo Reports Record Annual Sales, Profits," *Gamasutra*, May, 7, 2009, http://www.gamasutra.com/php-bin/news_index.php?story=23521.

10. Nintendo DS advertisement for *Brain Age*, back cover of *Discover* magazine, Spring 2007.

11. Sharon Begley, "How the Brain Rewires Itself," *Time*, January 29, 2007, 79.

12. Anya Martin, "Working with an Open Mind: Brain-Fitness Games Join Workplace, as Well as Senior Center, Arsenals," *MarketWatch*, May 13, 2009, http://www.marketwatch.com/story/brain-fitness-moves-from-senior-centers-to-the-job (accessed April 5, 2010).

13. Quoted in ibid.

14. Pam Belluck, "As Minds Age, What's Next? Brain Calisthenics," *New York Times*, December 27, 2006, http://www.nytimes.com/2006/12/27/health/27brain.html?_r=1 (accessed April 5, 2010).

15. Michel Foucault, "Governmentality," in *The Foucault Effect: Studies in Governmentality*, ed. Graham Burchell, Colin Gordon, and Pete Miller (Chicago: University of Chicago Press, 1991), 87–104. See also Majia Holmer Nadesan, *Governmentality, Biopower, and Everyday Life* (New York: Routledge, 2008).

16. See also Nikolas Rose, *Powers of Freedom: Reframing Political Thought* (Cambridge: Cambridge University Press, 1999); and Rose, *The Politics of Life Itself: Biomedicine, Power, and Subjectivity in the Twenty-First Century* (Princeton, NJ: Princeton University Press, 2007).

17. Deborah Cameron, "Who's Playing Who?," *The Age*, May 13, 2006, http://www.devin.com.au/devin-articles/2006/5/13/whos-playing-who/ (accessed April 5, 2010).

18. Nintendo DS *Brain Age* website, http://www.brainage.com/launch/training.jsp (accessed April 5, 2010).

19. Robert Crawford, "Healthism and the Medicalization of Everyday Life," *International Journal of Health Services* 10, no. 3 (1980): 380–381.

20. Leonard R. Kass, "Letter of Transmittal to the President," in *Beyond Therapy: Biotechnology and the Pursuit of Happiness*, by the President's Council on Bioethics (Washington, DC: President's Council on Bioethics, 2003), xv.

21. Ibid., xvii.

22. President's Council on Bioethics, *Beyond Therapy*, 261.

23. Kass, "Letter of Transmittal," xv.

24. President's Council on Bioethics, *Beyond Therapy*, 270.

25. President's New Freedom Commission on Mental Health, *Achieving the Promise: Transforming Mental Health Care in America* (Washington, DC: President's New Freedom Commission on Mental Health, 2003), 1.

26. Ibid., 2.

27. Ibid., 12.

28. For instance, see Peter Conrad and Joseph Schneider, *Deviance and Medicalization: From Badness to Sickness*, 2nd ed. (Philadelphia: Temple University Press, 1992); and Roy Porter, *Madness: A Brief History* (Oxford: Oxford University Press, 2002).

29. Rose, *Politics of Life Itself*. For more on medical imaging and new conceptions of the body, see also Lisa Cartwright, *Screening the Body: Tracing Medicine's Visual*

Culture (Minneapolis: University of Minnesota Press, 1995); Bettyann Kevles, *Naked to the Bone: Medical Imaging in the Twentieth Century* (New Brunswick, NJ: Rutgers University Press, 1997); Barbara Maria Stafford, *Body Criticism: Imaging the Unseen in Enlightenment Art and Medicine* (Cambridge, MA: MIT Press, 1991); and Catherine Waldby, *The Visible Human Project: Informatic Bodies and Posthuman Medicine* (New York: Routledge, 2000).

30. Richard Restak, *The New Brain: How the Modern Age Is Rewiring Your Mind* (New York: Rodale, 2003), 9.

31. "New Ways of Seeing the Brain," *Newsweek*, October 2009, http://photo.news week.com/content/photo/2009/10/new-ways-of-seeing-the-brain-photos.html (accessed April 20, 2010).

32. Stanley Finger, *Origins of Neuroscience: A History of Explanations into Brain Function* (New York: Oxford University Press, 1994); William Uttal, *The New Phrenology: The Limits of Localizing Cognitive Processes in the Brain* (Cambridge, MA: MIT Press, 2001); and Uttal, *Distributed Neural Systems: Beyond the New Phrenology* (Cornwall-on-Hudson, NY: Sloan Publishing, 2009). For more on the history of this controversy in brain science, see Anne Harrington, *Medicine, Mind, and the Double Brain* (Princeton, NJ: Princeton University Press, 1987).

33. Norman Doidge, *The Brain That Changes Itself* (New York: Penguin, 2007), 4, 12.

34. Uttal, *Distributed Neural Systems*.

35. Doidge, *Brain That Changes Itself*, 9.

36. Kass, "Letter of Transmittal," xvii.

37. Johnson, *Mind Wide Open*, 140–141.

38. David Healy, *The Antidepressant Era* (Cambridge, MA: Harvard University Press, 1997), 5.

39. I use neoliberal and neoliberalism as shorthand for a number of social changes, including the proliferation of entrepreneurial and market models to virtually every domain of existence, an increased emphasis on self-governance and respon-sibilization for risk, and widespread dissemination of optimization discourses to frame individuals' projects for living well. Rose describes neoliberalism as "a men-tality of government" that is tightly linked to the language of enterprise, where enterprise designates not only a form of organization but also "an image of a mode of activity to be encouraged in a multitude of arenas of life." *Inventing Ourselves: Psychology, Power, and Personhood* (Cambridge: Cambridge University Press, 1998), 153–154. David Harvey, quoted by Jodi Dean, defines neoliberalism as the endeavor "to bring all human action into the domain of the market," and Dean describes the term as designating "a philosophy viewing market exchange as a guide for all human action." See Dean, *Democracy and Other Neoliberal Fantasies* (Durham, NC: Duke University Press, 2009), 23, 51.

40. See, for example, Melinda Cooper, *Life as Surplus: Biotechnology and Capitalism in the Neoliberal Era* (Seattle: University of Washington Press, 2008); K. S. Rajan, *Biocapital: The Constitution of Postgenomic Life* (Durham, NC: Duke University Press, 2006); and Rose, *Politics of Life Itself*.

41. Daniel Amen, *Change Your Brain, Change Your Body* (New York: Harmony Books, 2010), 19.

42. Richard Restak, *Think Smart: A Neuroscientist's Prescription for Improving Your Brain's Performance* (New York: Riverhead Books, 2009), 21, 237.

43. I use control in the sense of Deleuze's "society of control." See Gilles Deleuze, "Postscript on the Societies of Control," in *Negotiations, 1972–1990*, trans. Martin Joughin (New York: Columbia University Press, 1995), 177–182.

44. Michel Foucault, *Discipline and Punish*, trans. Alan Sheridan (New York: Vintage Books, 1977), 137.

45. Michael Hardt and Antonio Negri, *Empire* (Cambridge, MA: Harvard University Press, 2000), 23–24, 330–331.

46. Deleuze, "Postscript," 178.

47. Jeffrey Nealon, *Foucault beyond Foucault* (Stanford, CA: Stanford University Press, 2008).

48. Hardt and Negri, *Empire*, 330–331.

49. Michel Foucault, *Security, Territory, Population*, trans. Graham Burchell (New York: Palgrave Macmillan, 2007), 19.

50. Ibid., 56–59.

51. Ibid, 57.

52. Deleuze, "Postscript," 179.

53. Monica Greco, "Psychosomatic Subjects and 'The Duty to Be Well': Personal Agency within Medical Rationality," *Economy and Society* 22, no. 3 (1993): 357–372.

54. Elkhonon Goldberg, *The New Executive Brain: Frontal Lobes in a Complex World* (New York: Oxford University Press, 2009), 109–110.

55. Restak, *New Brain*, 13 (emphasis mine).

56. Peter Kramer, *Listening to Prozac*, rev. ed. (New York: Penguin, 1997).

57. Paul Rabinow, "Artificiality and Enlightenment: From Sociobiology to Biosociality," in *Incorporations*, ed. Jonathan Crary and Sanford Kwinter (New York: Zone Books, 1992), 242.

58. Rose, *Politics of Life Itself*, 19.

59. Goldberg, *New Executive Brain*, 125.

60. Restak, *New Brain*, 25.

61. Goldberg, *New Executive Brain*, 125.

62. Susanne Antonetta, *A Mind Apart: Travels in a Neurodiverse World* (New York: Tarcher, 2005); and Amy Harmon, "Neurodiversity Forever: The Disability Movement Turns to Brains," *New York Times*, May 9, 2004, http://www.nytimes .com/2004/05/09/weekinreview/neurodiversity-forever-the-disability-movement- turns-to-brains.html (accessed June 13, 2007).

63. Rose, *Powers of Freedom*, 88.

64. Celeste Condit, *The Meanings of the Gene* (Madison: University of Wisconsin Press, 1999).

65. Ibid., 250.

66. Nikolas Rose uses the term "assemblage," taking it from Deleuze. For Rose, assemblages are "the localization and connecting together of routines, habits, and techniques within specific domains of action and value: libraries and studies, bedrooms and bathhouses, courtrooms and schoolrooms, consulting rooms and museum galleries, markets and department stores." *Inventing Ourselves*, 38. Similarly, Peter Miller and Nikolas Rose describe an assemblage as consisting of "diverse components, persons, forms of knowledge, technical procedure, and

modes of judgment and sanction." Each assemblage "is full of parts that come from elsewhere, strange couplings, chance relations, cogs, levers . . ." *Governing the Present* (Malden, MA: Polity, 2008), 200. The notion of assemblage is useful in foregrounding the fact that the rhetorical formations I trace are not "naturally" coherent but become connected through rhetoric, or through the circulation of common patterns of discourse.

67. Rose, *Inventing Ourselves*, 2.

68. Restak, *New Brain*.

CHAPTER 2   VISUALIZING THE NEW BRAIN

1. Richard Restak, *The Modular Brain* (New York: Touchstone, 1994), 1.

2. Richard Restak, *The New Brain: How the Modern Age Is Rewiring Your Mind* (New York: Rodale, 2003), 3.

3. Quoted in Susan Leigh Star, *Regions of the Mind* (Stanford, CA: Stanford University Press, 1989), xvii.

4. Restak, *New Brain*, 111.

5. Bruno Latour, "Drawing Things Together," in *Representation in Scientific Practice*, ed. Michael Lynch and Steve Woolgar (Cambridge, MA: MIT Press, 1990), 19–68.

6. Michel Foucault, introduction to *The Normal and the Pathological*, by Georges Canguilhem, trans. Carolyn R. Fawcett (New York: Zone Books, 1991), 14–15.

7. Discussions of articulation theory include Ronald Greene, "Another Materialist Rhetoric," *Critical Studies in Media Communication* 15, no. 1 (1998): 21–41; Nathan Stormer, "Articulation: A Working Paper on Rhetoric and *Taxis*," *Quarterly Journal of Speech* 90, no. 3 (2004): 257–284; Ernesto Laclau and Chantal Mouffe, *Hegemony and Socialist Strategy: Towards a Radical Democratic Politics* (New York: Verso, 1985); and Kevin DeLuca, "Articulation Theory: A Discursive Grounding for Rhetorical Practice," *Philosophy and Rhetoric* 32, no. 4 (1999): 334–348.

8. This idea has much in common with theories of coproduction that are being worked out in science and technology studies. Coproduction, as Sheila Jasanoff explains, is an "idiom" that has as its fundamental proposition "the ways in which we know and represent the world (both nature and society) are inseparable from the ways in which we choose to live in it." Scientific knowledge is not a "transcendent mirror of reality," but rather it "embeds and is embedded in social practices, identities, norms, conventions, discourses, instruments, and institutions." Coproduction deliberately tries to avoid both social and natural determinism and recognize the complexities of causal relationships. This view is influenced by Latour, who has argued that nature and culture are not a priori distinct domains, but rather are produced through patterns of discourse (including practice) that distribute causality in diverse ways to produce contingent relationships. See Jasanoff, "The Idiom of Co-production," in *States of Knowledge: The Co-production of Science and Social Order*, ed. Sheila Jasanoff (New York: Routledge, 2004), 2–3. See also Bruno Latour, *We Have Never Been Modern* (Cambridge, MA: Harvard University Press, 1993).

9. Quoted in Stanley Finger, *Origins of Neuroscience* (New York: Oxford, 1994), 32.

10. William Connolly draws attention to these patterns of circulation with his concept of "neuropolitics," the idea that culture, body, and brain mix together in

interactive networks. He describes the interaction of these networks as a "layered, tripartite assemblage" that is "marked by specific capacities of speed, reception, and enactment." Boundaries between nature and culture, or body and brain, are not static or ontological. Rather, forms and elements circulate among and between these contingent layers, interacting in what Connolly calls, following Deleuze, "zones of indiscernibility." See Connolly, *Neuropolitics: Thinking, Culture, Speed* (Minneapolis: University of Minnesota Press, 2002), 63.

11. Emily Martin offers an excellent discussion of how multiple arenas of biological and social life are understood as "systems," a framework that has personal, economic, and political consequences. Martin's discussion of systems has much in common with my usage of resonance. She argues that certain frameworks (e.g., systems or networks) come to "saturate" public discourse and popular understandings of the body and social life. In this sense, scientific understandings (such as those dominant in theories of the immune system or the nervous system) are imbricated in social and political changes. See *Flexible Bodies* (Boston: Beacon, 1994).

12. Michael Hardt and Antonio Negri, *Multitude: War and Democracy in the Age of Empire* (New York: Penguin, 2004), 330.

13. Ibid., 337.

14. Histories of neuroscience include Edwin Clarke and L. S. Jacyna, *Nineteenth-Century Origins of Neuroscientific Concepts* (Berkeley: University of California Press, 1987); Finger, *Origins of Neuroscience*; Lawrence McHenry, *Garrison's History of Neurology* (Springfield, IL: Charles C. Thomas Publishing, 1969); Star, *Regions of the Mind*; and Carl Zimmer, *Soul Made Flesh* (New York: Free Press, 2004).

15. While William Uttal in *The New Phrenology: The Limits of Localizing Cognitive Processes in the Brain* (Cambridge, MA: MIT Press, 2001), 11, states that this hypothesis is no longer a matter of open debate, recent books on plasticity cite evidence challenging even the localization of basic motor and sensory functions. See, for instance, Sharon Begley, *Train Your Mind, Change Your Brain* (New York: Ballantine, 2007); and Norman Doidge, *The Brain That Changes Itself* (New York: Penguin, 2007).

16. See Finger, *Origins of Neuroscience*; and Zimmer, *Soul Made Flesh*.

17. Finger, *Origins of Neuroscience*, 3.

18. This emphasis on smaller and smaller elements is conditioned by imaging technologies that visualize "life" at more minute levels. See Evelyn Fox Keller's chapter "The Visual Culture of Molecular Embryology" in *Making Sense of Life: Explaining Biological Development with Models, Metaphors, and Machines* (Cambridge, MA: Harvard University Press, 2002), 205–233. See also Nikolas Rose, *The Politics of Life Itself: Biomedicine, Power, and Subjectivity in the Twenty-First Century* (Princeton, NJ: Princeton University Press, 2007).

19. Uttal, *New Phrenology*, 26–27.

20. Ibid.

21. William Uttal, *Distributed Neural Systems: Beyond the New Phrenology* (Cornwall-on-Hudson, NY: Sloan Publishing, 2009), 103.

22. Ibid., 103.

23. Ibid.

24. Ibid., 142.

25. Ibid., 42.

26. Uttal, *New Phrenology*, 126.

27. Ibid., 145.

28. Ibid., 133–134.

29. See Melinda Beck, "When Anger Is an Illness," *Wall Street Journal*, March 9, 2010, http://online.wsj.com/article/SB10001424052748703954904575109671604585094 .html; Elizabeth Landau, "Revised Psychiatry Manual Targets Autism, Substance Disorders," *CNN Health*, February 10, 2010, http://www.cnn.com/2010/HEALTH/02/10/ dsm.v.revisions.psychiatry/index.html; Stuart Kirk and Herb Kutchins, *The Selling of DSM: The Rhetoric of Science in Psychiatry* (Hawthorne, NY: Aldine de Gruyter, 1992); and Paula Caplan, *They Say You're Crazy: How the World's Most Powerful Psychiatrists Decide Who's Normal* (Reading, MA: Addison-Wesley, 1995).

30. Rita Carter, *Mapping the Mind* (Berkeley: University of California Press, 1998), 8.

31. Jeffrey Schwartz and Sharon Begley, *The Mind and the Brain: Neuroplasticity and the Power of Mental Force* (New York: ReganBooks, 2002), 24.

32. Carter, *Mapping the Mind*, 11.

33. Robert Restak, *The Naked Brain: How the Emerging Neurosociety Is Changing How We Live, Work, and Love* (New York: Harmony Books, 2006), 214; and Elkhonon Goldberg, *The New Executive Brain* (New York: Oxford University Press, 2009), 37.

34. Charles Colbert, *A Measure of Perfection: Phrenology and the Fine Arts in America* (Chapel Hill: University of North Carolina Press, 1997), xiv.

35. Stephen Tomlinson, *Head Masters: Phrenology, Secular Education, and Nineteenth-Century Social Thought* (Tuscaloosa: University of Alabama Press, 2005), xi; and John von Wyhe, *Phrenology and the Origins of Victorian Naturalism* (Burlington, VT: Ashgate, 2004).

36. Colbert, *Measure of Perfection*, 23.

37. See Wyhe, *Phrenology and the Origins of Victorian Naturalism*, 11. Wyhe argues that phrenology was taken seriously as a science, and that phrenological naturalism was "one of the most influential ideological and cultural developments in Victorian Britain," a precursor to Darwin's theories. See also Madeleine Stern, *Heads and Headlines: The Phrenological Fowlers* (Norman: University of Oklahoma Press, 1971); and Tomlinson, *Head Masters*.

38. Wyhe, *Phrenology and the Origins of Victorian Naturalism*, 15.

39. Uttal, *New Phrenology*, 105.

40. Orson Squire Fowler and Lorenzo Niles Fowler, *Phrenology: A Practical Guide to Your Head* (New York: Chelsea House, 1980), 203.

41. Ibid., 2.

42. Colbert, *Measure of Perfection*, xii.

43. My discussion of the Fowlers follows most analyses, which read the Fowlers as using phrenology as a character-reading method. Michelle Gibbons has argued that most scholars neglect the extent to which the Fowlers believed that individuals could shape their brains and bodies through their choices and cultivation of habits. In other words, the Fowlers' phrenology was far more "plastic" than most historians assume. The Fowlers placed considerable emphasis on the malleability of children, but they also assumed that adults could change their bodies and brains through deliberate efforts. The Fowlers, Gibbons argues, offer

a far less deterministic account of brain-character and body-character correspondence than do the forefathers of phrenology (including Gall). Gibbons concludes,
"Reading character is certainly phrenology's dominant mode, however shaping
character and shaping the body is a significant corollary, particularly for the
immensely popular practical phrenology." See "The *American Phrenological Journal
and* Nineteenth-Century Visual Culture," paper presented at the AJHA-AEJMC
Joint Journalism Historians Conference, New York City, March 2006, 9.

44. Fowler and Fowler, *Phrenology*, 200–201.

45. Stern, *Heads and Headlines*, 37–38.

46. Andrew Norman, "Orson Fowler's Phrenology," in Fowler and Fowler, *Phrenology*, x.

47. Stern, *Heads and Headlines*, xiii.

48. Fowler and Fowler, *Phrenology*, 12.

49. Ibid., 189.

50. Gerald Edelman, *Bright Air, Brilliant Fire: On the Matter of the Mind* (New York: Basic Books, 1992), xiii.

51. Francis Crick, *The Astonishing Hypothesis: The Scientific Search for the Soul* (New York: Scribner, 1994), 259.

52. António Damásio, *Descartes' Error: Emotion, Reason, and the Human Brain* (New York: Avon, 1994), 226–227 (hereafter cited as *DE* in the text of this chapter).

53. See Daniel Dennett, *Consciousness Explained* (Boston: Little, Brown, 1991); and Dennett, *Freedom Evolves* (New York: Viking, 2003).

54. Uttal, *New Phrenology*, 13.

55. Mariana Valverde, *Diseases of the Will: Alcohol and the Dilemma of Freedom* (New York: Cambridge University Press, 1998).

56. Doidge, *Brain That Changes Itself*, 12.

57. Jeffrey Kluger, "What Makes Us Moral," *Time*, December 3, 2007, 60.

58. Carter, *Mapping the Mind*, 91.

59. Ibid. 92.

60. Noel Shafi, "Neuroscience and Law: The Evidentiary Value of Brain Imaging," *Graduate Student Journal of Psychology* 11 (2009): 27; Laura Stephens Khoshbin and Shahram Koshbin, "Imaging the Mind, Minding the Image: An Historical Introduction to Brain Imaging and the Law," *American Journal of Law and Medicine* 33, no. 2 (2007): 171–192; and Jeffrey Rosen, "The Brain on the Stand," *New York Times Magazine*, March 11, 2007, http://www.nytimes.com/2007/03/11/magazine/11Neurolaw.t.html.

61. Rosen, "Brain on the Stand."

62. Louann Brizendine, "Love, Sex, and the Male Brain," *CNN Opinion*, March 25, 2010, http://www.cnn.com/2010/OPINION/03/23/brizendine.male.brain/index.html.

63. From a rhetorical perspective, metaphors are not reducible to "literal" statements. In other words, metaphors are not simply accessories to discourse; rather, they constitute a fundamental activity of articulation and are essential components of scientific understandings of the world. See Celeste Condit, *The Meanings of the Gene* (Madison: University of Wisconsin Press, 1999); Donna Haraway, *Crystals, Fabrics, and Fields: Metaphors That Shape Embryos* (Berkeley, CA: North Atlantic Books, 2004); Evelyn Fox Keller, *Making Sense of Life*; Elizabeth Parthenia Shea, *How the Gene Got Its Groove: Figurative Language, Science, and the Rhetoric of the Real* (Albany:

State University of New York Press, 2008); and Steven Montgomery, *The Scientific Voice* (New York: Guilford, 1996).

64. If Damásio's version of the "New Brain" is not identical in form or function to the Cartesian brain, it also differs from the Cartesian mind. While the New Brain does take on certain features similar to those of the Cartesian mind, I do not believe that contemporary neuroscience is, as Elizabeth Wilson argues, "simply Cartesianism in material form." *Neural Geographies: Feminism and the Microstructure of Cognition* (New York: Routledge, 1998), 124. The brain-body division is not identical to the former arrangement of brain and mind—it looks different (literally, it is accompanied by and conceived via unique visualizing technologies), and it does different work. It enables different relations between nature and culture, different patterns of social organization, and different modes of subjectivation. The New Brain, unlike the Cartesian mind, can be visualized, theorized as "matter," and subjected to different interventions. And the New Brain, unlike the Cartesian brain, is not mechanistic, but rather is plastic and capable of alteration through drugs and calibration through images. To reduce it to an identity with Cartesianism risks missing its singularity, or the specificity of the effects that particular configurations of the brain produce.

65. Elizabeth Wilson, *Neural Geographies: Feminism and the Microstructure of Cognition* (New York: Routledge, 1998), 124.

66. Patrick R. Hof et al., "Cellular Components of Nervous Tissue," in *From Molecules to Networks: An Introduction to Cellular and Molecular Neuroscience*, ed. John H. Byrne and James L. Roberts (New York: Elsevier Academic Press, 2004), 1.

67. Debra Niehoff, *The Language of Life: How Cells Communicate in Health and Disease* (Washington, DC: Joseph Henry Press, 2005), 4.

68. Ibid., 189.

69. See especially Zimmer, *Soul Made Flesh*.

70. Francis Crick, *The Astonishing Hypothesis: The Scientific Search for the Soul* (New York: Charles Scribner's Sons, 1994), 268.

71. Edelman, *Bright Air, Brilliant Fire*, 8.

72. Ibid., 7 (emphasis mine).

73. Crick, *Astonishing Hypothesis*, xii, 4.

74. Quoted in Star, *Regions of the Mind*, 1.

75. Michael Gazzaniga, Richard Ivry, and George Mangun, *Cognitive Neuroscience: The Biology of the Mind* (New York: Norton, 1998), 11.

76. Daniel Goleman, preface to Begley, *Train Your Mind*, xii.

77. Goldberg, *New Executive Brain*, 47.

78. As Uttal explains, this shift is in part due to increasing recognition that when brain images are produced from digital data, the decision over what activity counts as data and what counts as noise is arbitrary. These arbitrary decisions determine what will show up as colored activity on the scan and what will not. Depending on where the data/noise threshold is set, images can vary significantly even for the same data set. Thus scientists are recognizing that more areas of the brain could be active than those reflected on the scan—some activity could be excluded from the scan simply because it is counted as "noise." See Uttal, *Distributed Neural Systems*, for an in-depth explanation.

79. Begley, *Train Your Mind, Change Your Brain* (hereafter cited as *TYMCYB* in the text of this chapter).

80. William R. Uttal, *The New Phrenology* (Cambridge, MA: MIT Press, 2001), 215.

81. Daniel Pink, *Free Agent Nation: The Future of Working for Yourself* (New York: Warner Business Books, 2001), 17.

82. Ibid., 16–17.

83. Ibid., 12.

84. See Ulrich Beck, *The Brave New World of Work* (Malden, MA: Polity, 2000); Jodi Dean, *Democracy and Other Neoliberal Fantasies* (Durham, NC: Duke University Press, 2009); and Jacques Donzelot, "Pleasure in Work," in *The Foucault Effect*, ed. Graham Burchell, Colin Gordon, and Peter Miller (Chicago: University of Chicago Press, 1991), 251–280.

85. Richard Sennett, *The Corrosion of Character: The Personal Consequences of Work in the New Capitalism* (New York: Norton, 1998), 9.

86. Michael Hammer and James Champy, *Reengineering the Corporation: A Manifesto for Business Revolution* (New York: HarperCollins, 1993), 68–70.

CHAPTER 3   PRACTICAL NEUROSCIENCE
AND BRAIN-BASED SELF-HELP

1. Daniel Amen, "SPECT Scans Offer Hope and Affirmation," *Dr. Amen's Blog*, February 1, 2010, http://www.amenclinics.com/blog/2689/spect-scans-offer-hope-and-affirmation/ (accessed April 3, 2010).

2. Daniel Amen, *Change Your Brain, Change Your Life* (New York: Times Books, 1998), 6.

3. Daniel Amen, *Healing the Hardware of the Soul: Enhance Your Brain to Improve Your Work, Love, and Spiritual Life* (New York: Free Press, 2002), 4.

4. Ibid., 5.

5. Amen, *Change Your Brain*, 228–229.

6. For further reading on therapy and governance, see Eva Illouz, *Saving the Modern Soul: Therapy, Emotions, and the Culture of Self-Help* (Berkeley: University of California Press, 2008); and Nikolas Rose, *Governing the Soul: The Shaping of the Private Self*, 2nd ed. (London: Free Association Books, 1999).

7. Daniel Amen, *Sex on the Brain: 12 Lessons to Enhance Your Love Life* (New York: Harmony Books, 2007), 90.

8. Ibid., 94.

9. Judy Quinn, "'Get a 'Life': The Year Begins with Self-Help Successes," *Publishers Weekly*, March 1, 1999, 25–26.

10. See, for instance, Harriet Hall, "A Skeptical View of SPECT Scans and Dr. Daniel Amen," *Quackwatch*, November 15, 2007, http://www.quackwatch.org/06ResearchProjects/amen.html (accessed April 14, 2010); and Robert Burton, "Brain Scam: Why Is PBS Airing Dr. Daniel Amen's Self-Produced Infomercial for the Prevention of Alzheimer's Disease?," Salon.com, May 12, 2008, http://www.salon.com/life/mind_reader/2008/05/12/daniel_amen (accessed April 14, 2010).

11. Quoted in Steve Salerno, *Sham: How the Self-Help Movement Made America Helpless* (New York: Crown, 2005), 9.

12. Ibid.

13. Micki McGee, *Self-Help, Inc.: Makeover Culture in American Life* (Oxford: Oxford University Press, 2005), 11.

14. Sandra Dolby, *Self-Help Books: Why Americans Keep Reading Them* (Urbana: University of Illinois Press, 2005).

15. Paul Rabinow, "Artificiality and Enlightenment: From Sociobiology to Biosociality," in *Incorporations*, ed. Jonathan Crary and Sanford Kwinter (New York: Zone Books, 1992), 242.

16. See Ian Hacking, "Self-Improvement," in *Foucault: A Critical Reader*, ed. D. C. Hoy (Oxford: Basil Blackwell, 1986), 235–241.

17. Daniel Amen, *Making a Good Brain Great: The Amen Clinic Program for Achieving and Sustaining Optimal Mental Performance* (New York: Harmony Books, 2005), 89–90 (hereafter cited as *MGBG* in text of this chapter).

18. Amen, *Change Your Brain*, 9.

19. Amen acknowledges this skepticism throughout his work, noting that his clinical applications are not widely accepted in the broader medical community.

20. Nikolas Rose, *The Politics of Life Itself: Biomedicine, Power, and Subjectivity in the Twenty-First Century* (Princeton, NJ: Princeton University Press, 2007), 140.

21. Ibid., 141.

22. Amen, *Change Your Brain*, 165.

23. Ibid., 177.

24. Joseph Dumit, *Picturing Personhood: Brain Scans and Biomedical Identity* (Princeton, NJ: Princeton University Press, 2004), 6.

25. Amen, *Change Your Brain*, 14.

26. Dana Cloud, *Control and Consolation in American Culture and Politics: Rhetoric of Therapy* (Thousand Oaks, CA: Sage, 1998).

27. Mariana Valverde, *Diseases of the Will: Alcohol and the Dilemmas of Freedom* (New York: Cambridge University Press, 1998), 64–65.

28. Amen, *Change Your Brain*, 49.

29. Nikolas Rose, *Inventing Ourselves: Psychology, Power, and Personhood* (Cambridge: Cambridge University Press, 1998), 17.

CHAPTER 4   BABIES, BLANK SLATES, AND BRAIN BUILDING

1. Quoted in Sarah Moughty, "The Zero-to-Three Debate: A Cautionary Look at Turning Science into Policy," *Frontline*, n.d., http://www.pbs.org/wgbh/pages/frontline/shows/teenbrain/science/zero.html (accessed December 19, 2005).

2. For more discussion of the Mozart initiatives and their tenuous scientific basis, see Adrian Bangerter and Chip Heath, "The Mozart Effect: Tracking the Evolution of a Scientific Legend," *British Journal of Social Psychology* 43, no. 3 (2004): 605–623.

3. For example, see Kathy Hirsh Pasek and Roberta Michnick Golinkoff, *Einstein Never Used Flashcards* (New York: Rodale, 2003); Pamela Paul, *Parenting, Inc.: How the Billion-Dollar Baby Business Has Changed the Way We Raise Our Children* (New York: Times Books, 2008); Juliet Schor, *Born to Buy: The Commercialized Child and the New Consumer Culture* (New York: Scribner, 2004); and Susan Gregory Thomas, *Buy, Buy Baby: How Consumer Culture Manipulates Parents and Harms Young Minds* (Boston: Houghton Mifflin, 2007). Despite these books' critical approach to consumerism

in child rearing, for the most part they tend to support the basic premises of the baby-brain movement: that children need particular attention during their early years to ensure proper neural development. The continued production of baby-brain books that emphasize emotional bonding and nurturing care further attests to the continuing power of the baby-brain discourses. See, for instance, Jill Stamm, *Bright from the Start: The Simple, Science-Backed Way to Nurture Your Child's Developing Mind from Birth to Age Three* (New York: Gotham Books, 2007); and Sue Gearhardt, *Why Love Matters: How Affection Shapes a Baby's Brain* (New York: Routledge, 2004).

4. Quoted in John Bruer, *The Myth of the First Three Years* (New York: Free Press, 1997), 8 (emphasis mine).

5. Ibid.

6. Beth Frerking, "Scientists Stress Need to Aid Parents," *Cleveland Plain Dealer*, April 18, 1997, 12A.

7. Society for Neuroscience, "White House Conference Melds Neuroscience and Public Policy," July/August 1997, http://web.sfn.org/NL/1997/July-August/white_house.html (accessed April 13, 2008).

8. Laura Bush, "Mrs. Bush's Remarks at the White House Summit on Early Childhood Cognitive Development," July 26, 2001, http://georgewbush-whitehouse.archives.gov/firstlady/news-speeches/speeches/fl20010726.html (accessed November 3, 2010).

9. Edward Zigler, Matia Finn-Stevenson, and Nancy Hall, *The First Three Years and Beyond: Brain Development and Social Policy* (New Haven, CT: Yale University Press, 2002), 7–8.

10. Ibid., 2.

11. Ibid., 3.

12. For a more in-depth history of scientific advice regarding early childhood, especially as it has affected mothers, see Rima Apple, *Perfect Motherhood: Science and Childrearing in America* (New Brunswick, NJ: Rutgers University Press, 2006); and Maxine Margolis, *True to Her Nature: Changing Advice to American Women* (Prospect Heights, IL: Waveland Press, 2000). For an example of an earlier instance of zero-to-three rhetoric, see Burton White, *The First Three Years of Life* (Englewood Cliffs, NJ: Prentice Hall, 1975).

13. Carnegie Corporation of New York, *Starting Points: Meeting the Needs of Our Youngest Children* (New York: Carnegie Corporation of New York, 1994), http://carnegie.org/fileadmin/Media/Publications/PDF/Starting%20Points%20Meeting%20the%20Needs%20of%20our%20Youngest%20Children.pdf (accessed April 20, 2010).

14. Wendy Cole, "Lighten Up, Folks: Are You Piping Mozart into the Crib and Reading Dr. Seuss around the Clock? Relax: To Develop a Baby's Brain, More Input Isn't Always Better," *Time*, October 19, 1998, 88–89.

15. Quoted in Bruer, *Myth of the First Three Years*, 52.

16. Zigler, Finn-Stevenson, and Hall, *First Three Years*, 193.

17. Bruer, *Myth of the First Three Years*, 52.

18. Ibid., 53.

19. Ibid., 47.

20. Ibid., 48.

21. Ibid., 51.

22. Matthew Melmed, "Informed or Overloaded," Letter, *Newsweek*, August 19–September 5, 2005, 18.

23. Jeffrey Kluger and Alice Park, "The Quest for a Superkid," *Time*, April 30, 2001, 50–55.

24. Donna Haraway, *Crystals, Fabrics, and Fields: Metaphors That Shape Embryos* (Berkeley, CA: North Atlantic Books, 2004); R. J. Sternberg, *Metaphors of Mind: Conceptions of the Nature of Intelligence* (Cambridge: Cambridge University Press, 1990); and Celeste Condit, *The Meanings of the Gene* (Madison: University of Wisconsin Press, 1999).

25. Steven Montgomery, *The Scientific Voice* (New York: Guilford, 1996), 137.

26. Haraway, *Crystals, Fabrics, and Fields*, 17.

27. Sharon Begley, "Your Child's Brain," *Newsweek*, February 19, 1996, 56.

28. Steven Pinker, *The Blank Slate* (New York: Viking, 2002).

29. Madeleine Nash, "Fertile Minds," *Time*, February 3, 1997, 50.

30. Bruer, *Myth of the First Three Years*, 49–50.

31. Nash, "Fertile Minds," 53.

32. Ibid., 48.

33. Pat Wingert and Anne Underwood, "First Steps: Hey—Look Out World, Here I Come," *Newsweek*, Special Issue, Spring/Summer 1997, 14.

34. LynNell Hancock and Pat Wingert, "The New Preschool," *Newsweek*, Special Issue, Spring/Summer 1997, 36–39.

35. Barbara Kantrowitz, "Off to a Good Start: Why the First Three Years Are So Crucial to a Child's Development," *Newsweek*, Special Issue, Spring/Summer 1997, 7.

36. Sharon Begley, "How to Build a Baby's Brain," Special Issue, *Newsweek*, Spring/Summer 1997, 28.

37. Ibid., 29–30.

38. Ibid., 30–31.

39. Debra Rosenberg, "Raising a Moral Child," *Newsweek*, Special Issue, Spring/Summer 1997, 92.

40. Begley, "Your Child's Brain," 58.

41. Ibid., 56.

42. Nash, "Fertile Minds," 55.

43. Ibid., 52.

44. Quoted in Begley, "How to Build a Baby's Brain," 32.

45. Nash, "Fertile Minds," 51; and Begley, "How to Build a Baby's Brain," 32.

46. Begley, "How to Build a Baby's Brain," 32.

47. "Growing Up Step By Step," *Newsweek*, Special Issue, Spring/Summer 1997, 26.

48. Ibid., 27.

49. "Milestones," *Newsweek*, August 15, 2005, 36.

50. Ibid., 37.

51. Carnegie Corporation, *Starting Points*.

52. Nikolas Rose, "Governing 'Advanced' Liberal Democracies," in *Foucault and Political Reason*, ed. Andrew Barry, Thomas Osborne, and Nikolas Rose (Chicago: University of Chicago Press, 1996), 37.

53. James Collins et al., "The Day Care Dilemma," *Time*, February 3, 1997, 60.

54. Ibid., 62.

55. B. Salzman, J. Carberry, and K. Hall, "Letter," *Newsweek*, August 29–September 5, 2005, 18.

56. Jack Bratich, Jeremy Packer, and Cameron McCarthy, "Governing the Present," in *Foucault, Cultural Studies, and Governmentality*, ed. Jack Bratich, Jeremy Packer, and Cameron McCarthy (Albany: State University of New York Press, 2003), 5.

CHAPTER 5   PILLS, POWER, AND THE NEUROSCIENCE OF EVERYDAY LIFE

1. See CHAADA website, http://www.chaada.org/ (accessed April 22, 2010).

2. See Fight for Kids website, http://www.fightforkids.org/ (accessed April 22, 2010).

3. Melody Petersen, *Our Daily Meds: How the Pharmaceutical Companies Transformed Themselves into Slick Marketing Machines and Hooked the Nation on Prescription Drugs* (New York: Picador, 2009); Marcia Angell, *The Truth about Drug Companies: How They Deceive Us and What to Do about It* (New York: Random House, 2005); and Timothy Scott, *America Fooled: The Truth about Antidepressants, Antipsychotics, and How We've Been Drugged* (Victoria, TX: Argo, 2006).

4. Gary Greenberg, *Manufacturing Depression: The Secret History of a Modern Disease* (New York: Simon & Schuster, 2010), 22.

5. Ibid., 13.

6. Emily Martin, *Flexible Bodies* (Boston: Beacon, 1994), 184.

7. Greenburg, *Manufacturing Depression*, 22.

8. As of November 2010, the exhibit was scheduled to appear at least through January 2011.

9. Critiques of the chemical imbalance theory include Robert Whitaker, *Anatomy of an Epidemic: Magic Bullets, Psychiatric Drugs, and the Astonishing Rise of Mental Illness in America* (New York: Crown, 2010); Irving Kirsch, *The Emperor's New Drugs: Exploding the Antidepressant Myth* (New York: Basic Books, 2010); and Joanna Moncrieff, *The Myth of the Chemical Cure: A Critique of Psychiatric Drug Treatment*, rev. ed. (New York: Palgrave Macmillan, 2009).

10. See Lawrence Rubin, ed., *Psychotropic Drugs and Popular Culture: Essays on Medicine, Mental Health, and the Media* (Jefferson, NC: McFarland, 2006).

11. Nikolas Rose, "Neurochemical Selves," *Society* 41, no. 1 (2003): 46–59.

12. Whitaker, *Anatomy of an Epidemic*, 3.

13. See Paul Rodriguez, "Talking Brains: A Cognitive Semantic Analysis of an Emerging Folk Neuropsychology," *Public Understanding of Science* 15, no. 3 (2006): 301–330; and Wolfgang Wagner, "Vernacular Science Knowledge: Its Role in Everyday Life Communication," *Public Understanding of Science* 16, no. 1 (2007): 7–22.

14. A thorough consideration of the anti-psychiatry and anti-medication discourses of recent decades is beyond the scope of this chapter. Sustained consideration of these discourses is a worthwhile project, and it would be especially valuable to study the extent to which these discourses reinforce the fundamental tenets of popular neuroscience despite their oppositional tone. For instance, some critiques of drugs rely on biomedical paradigms, contesting the chemical imbalance

theory of brain function but in the main supporting brain-based perceptions of human life. Again, while a full consideration is beyond the scope of this project, I suspect that some of these oppositional discourses ultimately perpetuate the brain-based terminologies and thinking patterns that have conditioned the spread of pharmaceuticals. Moreover, the tendency of these discourses to critique corporate deception is a limitation, as it neglects the broader social currents that provide corporate marketing with such a receptive audience for its pro-drug messages.

15. Eilean Hooper-Greenhill, *Museums and the Interpretation of Visual Culture* (New York: Routledge, 2000), x.

16. See, for example, Donald Preziosi, *Brain of the Earth's Body: Art, Museums, and the Phantasms of Modernity* (Minneapolis: University of Minnesota Press, 2003); Timothy Luke, *Museum Politics* (Minneapolis: University of Minnesota Press, 2002); Benedict Anderson, *Imagined Communities*, rev. ed. (New York: Verso, 2006); and Barbara Kirshenblatt-Gimblett, *Destination Culture: Tourism, Museums, and Heritage* (Berkeley: University of California Press, 1998).

17. Sharon MacDonald, "Supermarket Science? Consumers and 'The Public Understanding of Science,'" in *The Politics of Display: Museums, Science, Culture*, ed. Sharon MacDonald (New York: Routledge, 1998), 118.

18. See Tony Bennett, *The Birth of the Museum* (New York: Routledge, 1995). See also Barbara Kirshenblatt-Gimblett, "Exhibitionary Complexes," in *Museum Frictions*, ed. Ivan Karp et al. (Durham, NC: Duke University Press, 2006), 35–45; and in the same volume, Tony Bennett, "Exhibition, Difference, and the Logic of Culture," 46–69.

19. Bennett, "Exhibition, Difference, and the Logic of Culture," 57.

20. Luke, *Museum Politics*, 224; and MacDonald, "Supermarket Science?," 126.

21. John Durant, "Introduction," in *Museums and the Public Understanding of Science*, ed. John Durant (London: Science Museum in association with the Committee on the Public Understanding of Science, 1992), 8.

22. Houston Health Museum website, http://www.thehealthmuseum.org/ (accessed May 26, 2007).

23. MacDonald, "Supermarket Science?," 123.

24. Pfizer, *Brain: The World Inside Your Head* virtual tour, http://www.pfizer.com/brain/etour.html (accessed May 26, 2007).

25. Nikolas Rose, *Inventing Ourselves: Psychology, Power, and Personhood* (Cambridge: Cambridge University Press, 1998), 73.

26. Andrew Barry, "On Interactivity: Consumers, Citizens, and Culture," in *The Politics of Display: Museums, Science, Culture*, ed. Sharon MacDonald (New York: Routledge, 1998), 100.

27. Evergreen Exhibitions, "Interactives in the Exhibit," 2005, http://www.evergreen exhibitions.com/en/art/?7 (accessed May 26, 2007).

28. See Michel Foucault, "Technologies of the Self," in *Technologies of the Self: A Seminar with Michel Foucault*, ed. Luther H. Martin, Huck Gutman, and Patrick H. Hutton (Amherst: University of Massachusetts Press, 1988), 16–49.

29. The success of interactives is contested. For instance, Heath, vom Lehn, and Osborne find that the forms of interaction likely to take place are "prescribed by

the exhibit" (94) and minimal in effect. Alan Gross argues that so-called interactives reinforce the "passivity" (13) of visitors. Although the impact of interactives is likely overstated by the exhibit's producers, *Brain* does feature a variety of opportunities for social interaction and encourages activity through its displays and its extensions into the home and school. See Christian Heath, Dirk vom Lehn, and Jonathon Osborne, "Interaction and Interactives: Collaboration and Participation with Computer-Based Exhibits," *Public Understanding of Science* 14, no. 1 (2005): 91–101; and Alan Gross, "The Roles of Rhetoric in the Public Understanding of Science," *Public Understanding of Science* 3, no. 1 (1994): 3–23.

30. Jose Bonner, "Changing Strategies in Science Education," *Science*, October 8, 2004, 228.

31. Barry, "On Interactivity," 102.

32. Monica Greco, "Psychosomatic Subjects and the 'Duty to Be Well': Personal Agency within Medical Rationality," *Economy and Society* 22, no. 3 (1993): 357–372.

33. Greco, "Psychosomatic Subjects," 357.

34. Ibid., 359.

35. See also Nikolas Rose, *The Politics of Life Itself: Biomedicine, Power, and Subjectivity in the Twenty-First Century* (Princeton, NJ: Princeton University Press, 2007).

36. See Steven Shapin, "Cordelia's Love: Credibility and the Social Studies of Science," *Perspectives on Science* 3, no. 3 (1995): 270.

37. Evergreen Exhibitions, "Brain: The World Inside Your Head," 2005, http://www.evergreenexhibitions.com/exhibits/brain/index.asp (accessed April 22, 2010).

38. Rose, "Neurochemical Selves," 57.

39. Generally, the interiorization of psychiatric vocabularies produces subjects who think and act differently than subjects conditioned, for example, to a psychological vocabulary. The very nature of subjectivity is transformed. Graham Richards describes how the public uptake of psychotherapeutic vocabularies *created* people's experiences in very specific ways, channeling them to view and act according to terminologies such as "project, repress, regress" (85). Shapin writes of "the extent to which they [psychological vocabularies] are vernacularized" and "the extent to which they actually come to constitute the phenomenal base to which they refer." Human science can, he writes, regenerate and recreate "*human nature*" (267). See Graham Richards, *On Psychological Language and the Physiomorphic Basis of Human Nature* (New York: Routledge, 1989); and Shapin, "Cordelia's Love."

40. Evergreen Exhibitions. "Sponsor Benefits," http://www.evergreenexhibitions.com/sponsorship/index.asp (accessed February 26, 2007).

41. James Clifford uses the term "contact zones" for museums to describe their role as a relational space of ongoing contact rather than static collections. See *Routes: Travel and Translation in the Late Twentieth Century* (Cambridge, MA: Harvard University Press, 1997).

42. "Brain Basics Speaker Series" flyer, "Brain Fun Night" flyer, and "Excellent Field Trip Opportunity" flyer retrieved at the Houston Health Museum, March 11, 2007.

43. Mike Beirne and Sandra O'Loughlin, "Field Trips Appeal to Kids and Companies," *Brandweek*, May 15, 2006, 40.

44. Lynn Uyen Tran, "Teaching Science in Museums: The Pedagogy and Goal of Museum Educators," *Science Education* 91, no. 2 (2007): 278.

45. Martin Braund and Michael Reiss, "Towards a More Authentic Science Curriculum: The Contribution of Out-of-School Learning," *International Journal of Science Education* 28, no. 9 (2006): 1373–1388; and Doug Knapp, "A Longitudinal Analysis of an Out-of-School Science Experience," *School Science and Mathematics* 107, no. 2 (2007): 44–51.

46. Pfizer, *Brain: The World Inside Your Head Teacher's Activity Guide*, available by request from Pfizer and at http://www.pfizer.com/brain/teachers.html (accessed May 26, 2007).

47. Pamphlet retrieved at the Houston Health Museum, March 11 2007. Also available by request from Pfizer and at http://www.pfizer.com/brain/brochure.html (accessed March 26, 2007).

CHAPTER 6   MENTAL HEALTH CARE, THE RHETORIC OF RECOVERY, AND ENTREPRENEURIAL LIVES

1. A. Kathryn Power, "Mission Possible: Changing the Status Quo through System Transformation," address to the Florida Council for Community Mental Health Annual Conference, Clearwater Beach, FL, September 14, 2005.

2. Florida Self-Directed Care, http://flsdc.org/ (accessed April 30, 2010).

3. SAMHSA, *Free to Choose: Transforming Behavioral Health Care to Self-Direction* (Rockville, MD: Substance Abuse and Mental Health Services Administration, 2005), 11.

4. Tom K. J. Craig, "Recovery: Say What You Mean and Mean What You Say," *Journal of Mental Health* 17, no. 2 (2008): 125.

5. A. Kathryn Power, "Remarks by A. Kathryn Power," address to the National Consensus Conference on Mental Health and Systems Transformation, Rockville, MD, December 16, 2004.

6. For additional critical perspectives on the New Freedom Commission, see Toby Miller, *Makeover Nation: The United States of Reinvention* (Columbus: Ohio State University Press, 2008); and Nikolas Rose, *The Politics of Life Itself: Biomedicine, Power, and Subjectivity in the Twenty-First Century* (Princeton, NJ: Princeton University Press, 2007).

7. George W. Bush. "President Says U.S. Must Make Commitment to Mental Health Care," address at University of New Mexico, Continuing Education Center, Albuquerque, NM, April 29, 2002.

8. President's New Freedom Commission on Mental Health, *Achieving the Promise: Transforming Mental Health Care in America* (Washington, DC: President's New Freedom Commission on Mental Health, 2003), 5.

9. Michael Hogan, "The President's New Freedom Commission: Recommendations to Transform Mental Health Care in America," *Psychiatric Services* 54, no. 11 (2003): 146.

10. U.S. Department of Health and Human Services, *Mental Health: A Report of the Surgeon General—Executive Summary* (Rockville, MD: U.S. Department of Health and Human Services, Substance Abuse and Mental Health Services Administration, Center for Mental Health Services, National Institutes of Health, National Institute of Mental Health, 1999), 80.

11. New Freedom Commission, *Achieving the Promise*, 3–4.

12. *Olmstead v. L. C./E. W.*, 119 SCt 2176 (1999).

13. For instance, see Gerald Grob's history of mental health care in the United States, *Mental Illness and American Society, 1875–1940* (Princeton, NJ: Princeton University Press, 1983). In addition, see Mick Carpenter, "'It's a Small World': Mental Health Policy under Welfare Capitalism since 1945," *Sociology of Health and Illness* 22, no. 5 (2000): 602–620; and Jim Mansell, "Deinstiutionalisation and Community Living: Progress, Problems, and Priorities," *Journal of Intellectual and Developmental Disability* 31, no. 2 (2006): 65–76.

14. See Carpenter, "It's a Small World,"; Grob, *Mental Illness and American Society*; and Steven Gillon, *That's Not What We Meant to Do: Reform and Its Unintended Consequences in the Twentieth Century* (New York: Norton, 2000).

15. Grob, *Mental Illness and American Society*, 426.

16. Quoted in Gillon, *That's Not What We Meant*, 92.

17. Ibid., 91.

18. Gerald Grob, "Public Policy and Mental Illnesses: Jimmy Carter's Presidential Commission on Mental Health," *Milbank Quarterly* 83, no. 3 (2005): 427.

19. Ibid., 425–456.

20. Ibid., 427–428.

21. Ibid., 449.

22. Robert Bernstein, "Statement from the Bazelon Center for Mental Health Law," part of "The New Freedom Commission Report: The Campaign for Mental Health Reform: A New Advocacy Partnership," *Psychiatric Services* 54, no. 11 (2003): 1479.

23. "New Freedom Commission Report," 1475.

24. Ibid., 1476.

25. Michael J. Stoil, "The Aftermath of the New Freedom Commission," *Behavioral Health Management*, January/February 2004, 11–12.

26. New Freedom Commission, *Achieving the Promise*, 5.

27. Richard Birkel, "Statement from the National Alliance for the Mentally Ill," part of "New Freedom Commission Report," 1478.

28. George W. Bush, "Executive Order 13263 of April 29, 2002," appendix to New Freedom Commission, *Achieving the Promise: Transforming Mental Health Care in America*.

29. New Freedom Commission, *Achieving the Promise*, 57, 12, 67.

30. Ibid., 27.

31. Ibid., 29.

32. Depression and Bipolar Support Alliance, *The State of Depression in America* (Chicago: Depression and Bipolar Support Alliance, 2006), 1.

33. Ibid., 14.

34. New Freedom Commission, *Achieving the Promise*, 12.

35. "National Consensus Statement on Recovery," National Consensus Conference on Mental Health Recovery and Systems Transformation, December 16–17, 2004, http://www.samhsa.gov/news/newsreleases/060215_consumer.htm (accessed November 5, 2010).

36. Steven J. Onken et al., "An Analysis of the Definitions and Elements of Recovery: A Review of the Literature," *Psychiatric Rehabilitation Journal* 31, no. 1 (2007): 9–22.

37. New Freedom Commission, *Achieving the Promise*, 28.

38. Ibid., 8.

39. Thomas Nerney, "Quality Issues in Consumer/Family Direction," March 2004, 1–2. Available at http://mentalhealth.samhsa.gov/publications/allpubs/NMH05–0194/default.asp (accessed April 30, 2010).

40. Ibid., 6–7.

41. UPENN Collaborative on Community Integration, "Self-Directed Care in Public Mental Health Systems," n.d., http://www.upennrrtc.org/var/tool/file/137-CI%20Tool%20-%20Self-Directed%20Care.pdf (accessed April 30, 2010).

42. SAMHSA, *Free to Choose*, 9.

43. Michel Foucault, *The Birth of Biopolitics*, trans. Graham Burchell (New York: Palgrave Macmillan, 2008), 230.

44. Ibid., 242.

45. Jodi Dean, *Democracy and Other Neoliberal Fantasies* (Durham, NC: Duke University Press, 2009), 23.

CHAPTER 7   THE BRAIN IS THE FRONTIER

1. Rita Carter, *Mapping the Mind* (Berkeley: University of California Press, 1998), 8.

2. Quoted in Sharon Begley, "The Brain: Science Opens New Windows on the Mind," *Newsweek*, April 4, 1992, 66.

3. Bill Clinton, statement announcing the White House Conference on Early Childhood Development and Learning, March 13, 1997, http://www.ed.gov/PressReleases/04–1997/70417d.html (accessed February 19, 2006).

4. Zack Lynch with Byron Laursen, *The Neuro Revolution: How Brain Science Is Changing Our World* (New York: St. Martin's Press, 2009), 7–8.

5. Richard Restak, *The Naked Brain: How the Emerging Neurosociety Is Changing How We Live, Work, and Love* (New York: Harmony Books, 2006), 231.

6. Nancy Andreasen, *Brave New Brain: Conquering Mental Illness in the Era of the Genome* (Oxford: Oxford University Press, 2001), 4.

7. Lynch, *Neuro Revolution*, 11.

8. Rick Hanson, *Buddha's Brain: The Practical Neuroscience of Happiness, Love, and Wisdom* (Oakland, CA: New Harbinger Publications, 2009), 1.

9. Ibid., 18.

10. Peter Miller and Nikolas Rose, *Governing the Present* (Malden, MA: Polity, 2008), 100–101.

11. See Anjana Ahuja, "God on the Brain: Is Religion Just a Step away from Mental Illness?," *London Times*, April 17, 2003, http://www.timesonline.co.uk/tol/life_and_style/article1131051.ece (accessed October 30, 2004).

12. Jeffrey Kluger, "Is God in Our Genes?" *Time*, October 25, 2004, http://www.time.com/time/magazine/article/0,9171,995465–1,00.html (accessed October 31, 2004).

13. Andrew Newberg and Mark Waldman, *How God Changes Your Brain: Breakthrough Findings from a Leading Neuroscientist* (New York: Ballantine, 2009), 6–7.

14. Ibid., 14.

15. Ibid., 21.

16. Miller and Rose, *Governing the Present*, 100–101.

17. Frederick Jackson Turner, *The Frontier in American History* (1920; repr., Tucson: University of Arizona Press, 1986), 2–3.

18. Ibid., 4.

19. Ibid., 38.

20. Ibid.

21. Ibid., 65.

22. Ibid., 14–15.

23. Ibid., 269.

24. Gilles Deleuze, *The Logic of Sense*, trans. Mark Lester (New York: Columbia University Press, 1990), 222.

25. Ibid., 223.

26. E. O. Wilson, *Consilience: The Unity of Knowledge* (New York: Vintage, 1998), 106.

SELECTED BIBLIOGRAPHY

Amen, Daniel. *Change Your Brain, Change Your Body*. New York: Harmony Books, 2010.

———. *Change Your Brain, Change Your Life*. New York: Times Books, 1998.

———. *Healing the Hardware of the Soul: Enhance Your Brain to Improve Your Work, Love, and Spiritual Life*. New York: Free Press, 2002.

———. *Making a Good Brain Great: The Amen Clinic Program for Achieving and Sustaining Optimal Mental Performance*. New York: Harmony Books, 2005.

———. *Sex on the Brain: 12 Lessons to Enhance Your Love Life*. New York: Harmony Books, 2007.

Anderson, Benedict. *Imagined Communities*. Rev. ed. New York: Verso, 2006.

Andianopoulos, Georgia. *Retrain Your Brain, Reshape Your Body: The Breakthrough Brain-Changing Weight-Loss Program*. New York: McGraw-Hill, 2008.

Andreasen, Nancy. *Brave New Brain: Conquering Mental Illness in the Age of the Genome*. Oxford: Oxford University Press, 2001.

Angell, Marcia. *The Truth about Drug Companies: How They Deceive Us and What to Do about It*. New York: Random House, 2005.

Antonetta, Susanne. *A Mind Apart: Travels in a Neurodiverse World*. New York: Tarcher, 2005.

Apple, Rima. *Perfect Motherhood: Science and Childrearing in America*. New Brunswick, NJ: Rutgers University Press, 2006.

Arden, John. *Rewire Your Brain: Think Your Way to a Better Life*. Hoboken, NJ: Wiley, 2010.

Bangerter, Adrian, and Chip Heath. "The Mozart Effect: Tracking the Evolution of a Scientific Legend." *British Journal of Social Psychology* 43, no. 3 (2004): 605–623.

Barry, Andrew. "On Interactivity: Consumers, Citizens, and Culture." In *The Politics of Display: Museums, Science Culture*, edited by Sharon MacDonald, 98–117. New York: Routledge, 1998.

Beaulieu, Anne. "The Brain at the End of the Rainbow: The Promises of Brain Scans in the Research Field and in the Media." In *Wild Science: Reading Feminism, Medicine, and the Media*, edited by Janine Marchessault and Kim Sawchuck, 39–52. New York: Routledge, 2000.

Beck, Ulrich. *The Brave New World of Work*. Malden, MA: Polity, 2000.

Begley, Sharon. *Train Your Mind, Change Your Brain*. New York: Ballantine, 2007.

Bennett, Tony. *The Birth of the Museum*. New York: Routledge, 1995.

———. "Exhibition, Difference, and the Logic of Culture." In *Museum Frictions*, edited by Ivan Karp, Corinne A. Kratz, Lynn Szwaja, and Tomas Ybarra-Frausto, 46–69. Durham, NC: Duke University Press.

Bonner, Jose. "Changing Strategies on Science Education." *Science*, October 8, 2004.

Bratich, Jack, Jeremy Packer, and Cameron McCarthy. "Governing the Present." In *Foucault, Cultural Studies, and Governmentality*, edited by Jack Bratich, Jeremy Packer, and Cameron McCarthy, 3–22. Albany: State University of New York Press, 2003.

Braund, Martin, and Michael Reiss. "Towards a More Authentic Science Curriculum: The Contribution of Out-of-School Learning." *International Journal of Science Education* 28, no. 9 (2006): 1373–1388.

Bruer, John. *The Myth of the First Three Years*. New York: Free Press, 1997.

Byrne, John H., and James L. Roberts, eds. *From Molecules to Networks: An Introduction to Cellular and Molecular Neuroscience*. New York: Elsevier Academic Press, 2004.

Caplan, Paula. *They Say You're Crazy: How the World's Most Powerful Psychiatrists Decide Who's Normal*. Reading, MA: Addison-Wesley, 1995.

Carpenter, Mick. "'It's a Small World': Mental Health Policy under Welfare Capitalism since 1945." *Sociology of Health and Illness* 22, no. 5 (2000): 602–620.

Carter, Rita. *Mapping the Mind*. Berkeley: University of California Press, 1998.

Cartwright, Lisa. *Screening the Body: Tracing Medicine's Visual Culture*. Minneapolis: University of Minnesota Press, 1995.

Clarke, Edwin, and L. S. Jacyna. *Nineteenth-Century Origins of Neuroscientific Concepts*. Berkeley: University of California Press, 1987.

Clifford, James. *Routes: Travel and Translation in the Late Twentieth Century*. Cambridge, MA: Harvard University Press, 1997.

Cloud, Dana. *Control and Consolation in American Culture and Politics: Rhetoric of Therapy*. Thousand Oaks, CA: Sage, 1998.

Colbert, Charles. *A Measure of Perfection: Phrenology and the Fine Arts in America*. Chapel Hill: University of North Carolina Press, 1997.

Condit, Celeste. *The Meanings of the Gene*. Madison: University of Wisconsin Press, 1999.

Connolly, William. *Neuropolitics: Thinking, Culture, Speed*. Minneapolis: University of Minnesota Press, 2002.

Conrad, Peter, and Joseph Schneider. *Deviance and Medicalization: From Badness to Sickness*. 2nd ed. Philadelphia: Temple University Press, 1992.

Cooper, Melinda. *Life as Surplus: Biotechnology and Capitalism in the Neoliberal Era*. Seattle: University of Washington Press, 2008.

Craig, Tom K. J. "Recovery: Say What You Mean and Mean What You Say." *Journal of Mental Health* 17, no. 2 (2008): 125–128.

Crawford, Robert. "Healthism and the Medicalization of Everyday Life." *International Journal of Health Services* 10, no. 3 (1980): 365–388.

Crick, Francis. *The Astonishing Hypothesis: The Scientific Search for the Soul*. New York: Scribner, 1994.

Damásio, António. *Descartes' Error: Emotion, Reason, and the Human Brain*. New York: Avon, 1994.

Dean, Jodi. *Democracy and Other Neoliberal Fantasies*. Durham, NC: Duke University Press, 2009.

Deleuze, Gilles. *The Logic of Sense*. Translated by Mark Lester. New York: Columbia University Press, 1990.

———. "Postscript on the Societies of Control." In *Negotiations, 1972–1990*, translated by Martin Joughin, 177–182. New York: Columbia University Press, 1995.

DeLuca, Kevin. "Articulation Theory: A Discursive Grounding for Rhetorical Practice." *Philosophy and Rhetoric* 32, no. 4 (1999): 334–348.

Dennett, Daniel. *Consciousness Explained*. Boston: Little, Brown, 1991.

———. *Freedom Evolves*. New York: Viking, 2003.

Doidge, Norman. *The Brain That Changes Itself.* New York: Penguin Books, 2007.

Dolby, Sandra. *Self-Help Books: Why Americans Keep Reading Them.* Urbana: University of Illinois Press, 2005.

Donzelot, Jacques. "Pleasure in Work." In *The Foucault Effect*, edited by Graham Burchell, Colin Gordon, and Pete Miller, 251–280. Chicago: University of Chicago Press, 1991.

Dumit, Joseph. *Picturing Personhood: Brain Scans and Biomedical Identity.* Princeton, NJ: Princeton University Press, 2004.

Durant, John. "Introduction." In *Museums and the Public Understanding of Science*, edited by John Durant. London: Science Museum in association with the Committee on the Public Understanding of Science, 1992.

Edelman, Gerald. *Bright Air, Brilliant Fire: On the Matter of the Mind.* New York: Basic Books, 1992.

Finger, Stanley. *Origins of Neuroscience: A History of Explanations into Brain Function.* New York: Oxford University Press, 1994.

Foucault, Michel. *The Birth of Biopolitics.* Translated by Graham Burchell. New York: Palgrave Macmillan, 2008.

———. *Discipline and Punish.* Translated by Alan Sheridan. New York: Vintage Books, 1977.

———. "Governmentality." In *The Foucault Effect: Studies in Governmentality*, edited by Graham Burchell, Colin Gordon, and Pete Miller, 87–104. Chicago: University of Chicago Press, 1991.

———. Introduction to *The Normal and the Pathological*, by Georges Canguilhem, translated by Carolyn R. Fawcett, 7–24. New York: Zone Books, 1991.

———. *Security, Territory, Population.* Translated Graham Burchell. New York: Palgrave Macmillan, 2007.

———. "Technologies of the Self." In *Technologies of the Self: A Seminar with Michel Foucault*, edited by Luther H. Martin, Huck Gutman, and Patrick H. Hutton, 16–49. Amherst: University of Massachusetts Press, 1988.

Gazzaniga, Michael, Richard Ivry, and George Mangun. *Cognitive Neuroscience: The Biology of the Mind.* New York: Norton, 1998.

Gearhardt, Sue. *Why Love Matters: How Affection Shapes a Baby's Brain.* New York: Routledge, 2004.

Gibbons, Michelle. "Seeing the Brain in the Matter: Functional Brain Imaging as Framed Visual Argument." *Argumentation and Advocacy* 43, nos. 3/4 (2007): 175–188.

Gillon, Steven. *That's Not What We Meant to Do: Reform and Its Unintended Consequences in the Twentieth Century.* New York: Norton, 2000.

Goldberg, Elkhonon. *The New Executive Brain: Frontal Lobes in a Complex World.* New York: Oxford University Press, 2009.

Greco, Monica. "Psychosomatic Subjects and the 'Duty to Be Well': Personal Agency within Medical Rationality." *Economy and Society* 22, no. 3 (1993): 357–372.

Greenberg, Gary. *Manufacturing Depression: The Secret History of a Modern Disease.* New York: Simon & Schuster, 2010.

Greene, Ronald. "Another Materialist Rhetoric." *Critical Studies in Media Communication* 15, no. 1 (1998): 21–41.

Grob, Gerald. *Mental Illness and American Society, 1975–1940.* Princeton, NJ: Princeton University Press, 1983.

———. "Public Policy and Mental Illnesses: Jimmy Carter's Presidential Commission on Mental Health." *Milbank Quarterly* 83, no. 3 (2005): 425–456.

Gross, Alan. "The Roles of Rhetoric in the Public Understanding of Science." *Public Understanding of Science* 3, no. 1 (1994): 3–23.

Hacking, Ian. "Self-Improvement." In *Foucault: A Critical Reader*, edited by D. C. Hoy, 235–241. Oxford: Basil Blackwell, 1986.

Hammer, Michael, and James Champy. *Reengineering the Corporation: A Manifesto for Business Revolution*. New York: HarperCollins, 1993.

Hanson, Rick. *Buddha's Brain: The Practical Neuroscience of Happiness, Love, and Wisdom*. Oakland, CA: New Harbinger Publications, 2009.

Haraway, Donna. *Crystals, Fabrics, and Fields: Metaphors That Shape Embryos*. Berkeley, CA: North Atlantic Books, 2004.

Hardt, Michael, and Antonio Negri. *Empire*. Cambridge, MA: Harvard University Press, 2000.

———. *Multitude: War and Democracy in the Age of Empire*. New York: Penguin, 2004.

Harrington, Anne. *Medicine, Mind, and the Double Brain*. Princeton, NJ: Princeton University Press, 1987.

Healy, David. *The Antidepressant Era*. Cambridge, MA: Harvard University Press, 1997.

Heath, Christian, Dirk von Lehn, and Jonathon Osborne. "Interaction and Interactives: Collaboration and Participation with Computer-Based Exhibits." *Public Understanding of Science* 14, no. 1 (2005): 91–101.

Hogan, Michael. "The President's New Freedom Commission: Recommendations to Transform Mental Health Care in America." *Psychiatric Services* 54, no. 11 (2003): 1467–1474.

Hooper-Greenhill, Eilean. *Museums and the Interpretation of Visual Culture*. New York: Routledge, 2000.

Illouz, Eva. *Saving the Modern Soul: Therapy, Emotions, and the Culture of Self-Help*. Berkeley: University of California Press, 2008.

Jasanoff, Sheila. "The Idiom of Co-production." In *States of Knowledge: The Co-production of Science and Social Order*, edited by Sheila Jasanoff, 1–12. New York: Routledge, 2004.

Johnson, Steven. *Mind Wide Open: Your Brain and the Neuroscience of Everyday Life*. New York: Scribner, 2004.

Joyce, Kelly. *Magnetic Appeal: MRI and the Myth of Transparency*. Ithaca, NY: Cornell University Press, 2008.

Kawashima, Ryuta. *Train Your Brain More: Better Brainpower, Better Memory, Better Creativity*. New York: Penguin, 2008.

Keller, Evelyn Fox. *Making Sense of Life: Explaining Biological Development with Models, Metaphors, and Machines*. Cambridge, MA: Harvard University Press, 2002.

Kevles, Bettyann. *Naked to the Bone: Medical Imaging in the Twentieth Century*. New Brunswick, NJ: Rutgers University Press, 1997.

Kirk, Stuart, and Herb Kutchins. *The Selling of DSM: The Rhetoric of Science in Psychiatry*. Hawthorne, NY: Aldine de Gruyter, 1992.

Kirsch, Irving. *The Emperor's New Drugs: Exploding the Antidepressant Myth*. New York: Basic Books, 2010.

Kirshenblatt-Gimblett, Barbara. *Destination Culture: Tourism, Museums, and Heritage*. Berkeley: University of California Press, 1998.

———. "Exhibitionary Complexes." In *Museum Frictions*, edited by Ivan Karp, Corinne A. Kratz, Lynn Szwaja, and Tomas Ybarra-Frausto, 35–45. Durham, NC: Duke University Press, 2006.

Knapp, Doug. "A Longitudinal Analysis of an Out-of-School Science Experience." *School Science and Mathematics* 107, no. 2 (2007): 44–51.

Kramer, Peter. *Listening to Prozac*. Rev. ed. New York: Penguin, 1997.

Laclau, Ernesto, and Chantal Mouffe. *Hegemony and Socialist Strategy: Towards a Radical Democratic Politics*. New York: Verso, 1985.

Latour, Bruno. "Drawing Things Together." In *Representation and Scientific Practice*, edited by Michael Lynch and Stephen Woolgar, 19–68. Cambridge, MA: MIT Press, 1990.

———. *We Have Never Been Modern*. Cambridge, MA: Harvard University Press, 1993.

Luke, Timothy. *Museum Politics*. Minneapolis: University of Minnesota Press, 2002.

Lynch, Zack, with Byron Laursen. *The Neuro Revolution: How Brain Science Is Changing Our World*. New York: St. Martin's Press, 2009.

MacDonald, Sharon. "Supermarket Science? Consumers and 'The Public Understanding of Science.'" In *The Politics of Display: Museums, Science, Culture*, edited by Sharon MacDonald, 118–138. New York: Routledge, 1998.

Mansell, Jim. "Deinstitutionalisation and Community Living: Progress, Problems, and Priorities." *Journal of Intellectual and Developmental Disability* 31, no. 2 (2006): 65–76.

Margolis, Maxine. *True to Her Nature: Changing Advice to American Women*. Prospect Heights, IL: Waveland Press, 2000.

Martin, Emily. *Flexible Bodies*. Boston: Beacon, 1994.

McGee, Micki. *Self-Help, Inc.: Makeover Culture in American Life*. Oxford: Oxford University Press, 2005.

McHenry, Lawrence. *Garrison's History of Neurology*. Springfield, IL: Charles C. Thomas Publishing, 1969.

Miller, Peter, and Nikolas Rose. *Governing the Present*. Malden, MA: Polity, 2008.

Miller, Toby. *Makeover Nation: The United States of Reinvention*. Columbus: Ohio State University Press, 2008.

Moncrieff, Joanna. *The Myth of the Chemical Cure: A Critique of Psychiatric Drug Treatment*. Rev. ed. New York: Palgrave Macmillan, 2009.

Montgomery, Steven. *The Scientific Voice*. New York: Guilford, 1996.

Nadesan, Majia Holmer. *Governmentality, Biopower, and Everyday Life*. New York: Routledge, 2008.

Nealon, Jeffrey. *Foucault beyond Foucault*. Stanford, CA: Stanford University Press, 2008.

Newberg, Andrew, and Mark Waldman. *How God Changes Your Brain: Breakthrough Findings from a Leading Neuroscientist*. New York: Ballantine, 2009.

Niehoff, Debra. *The Language of Life: How Cells Communicate in Health and Illness*. Washington, DC: Joseph Henry Press, 2005.

Pasek, Kathy Hirsh, and Roberta Michnick Golinkoff. *Einstein Never Used Flashcards*. New York: Rodale, 2003.

Paul, Pamela. *Parenting, Inc.: How the Billion-Dollar Baby Business Has Changed the Way We Raise Our Children*. New York: Times Books, 2008.

Petersen, Melody. *Our Daily Meds: How the Pharmaceutical Companies Transformed Themselves into Slick Marketing Machines and Hooked the Nation on Prescription Drugs*. New York: Picador, 2009.

Pink, Daniel. *Free Agent Nation: The Future of Working For Yourself*. New York: Warner Business Books, 2001.

Pinker, Steven. *The Blank Slate*. New York: Viking, 2002.

Porter, Roy. *Madness: A Brief History*. Oxford: Oxford University Press, 2002.

Preziosi, Donald. *Brain of the Earth's Body: Art, Museums, and the Phantasms of Modernity*. Minneapolis: University of Minnesota Press, 2003.

Rabinow, Paul. "Artificiality and Enlightenment: From Sociobiology to Biosociality." In *Incorporations*, edited by Jonathan Crary and Sanford Kwinter, 234–252. New York: Zone Books, 1992.

Rajan, K. S. *Biocapital: The Constitution of Postgenomic Life.* Durham, NC: Duke University Press, 2006.

Restak, Richard. *The Modular Brain.* New York: Touchstone, 1994.

———. *The Naked Brain: How the Emerging Neurosociety Is Changing How We Live, Work, and Love.* New York: Harmony Books, 2006.

———. *The New Brain: How the Modern Age Is Rewiring Your Mind.* New York: Rodale, 2003.

———. *Think Smart: A Neuroscientist's Prescription for Improving Your Brain's Performance.* New York: Riverhead Books, 2009.

Richards, Graham. *On Psychological Language and the Physiomorphic Basis of Human Nature.* New York: Routledge, 1989.

Rodriguez, Paul. "Talking Brains: A Cognitive Semantic Analysis of an Emerging Folk Neuropsychology." *Public Understanding of Science* 15, no. 3 (2006): 301–330.

Rose, Nikolas. "Governing 'Advanced' Liberal Democracies." In *Foucault and Political Reason*, edited by Andrew Barry, Thomas Osborne, and Nikolas Rose, 37–64. Chicago: University of Chicago Press, 1996.

———. *Governing the Soul: The Shaping of the Private Self.* 2nd ed. London: Free Association Books, 1999.

———. *Inventing Ourselves: Psychology, Power, and Personhood.* Cambridge: Cambridge University Press, 1998.

———. "Neurochemical Selves." *Society* 41, no. 1 (2003): 46–59.

———. *The Politics of Life Itself: Biomedicine, Power, and Subjectivity in the Twenty-First Century.* Princeton, NJ: Princeton University Press, 2007.

———. *Powers of Freedom: Reframing Political Thought.* Cambridge: Cambridge University Press, 1999.

Rubin, Lawrence, ed. *Psychotropic Drugs and Popular Culture: Essays on Medicine, Mental Health, and the Media.* Jefferson, NC: McFarland, 2006.

Salerno, Steve. *Sham: How the Self-Help Movement Made America Helpless.* New York: Crown, 2005.

Schor, Juliet. *Born to Buy: The Commercialized Child and the New Consumer Culture.* New York: Scribner, 2004.

Schwartz, Jeffrey, and Sharon Begley. *The Mind and the Brain: Neuroplasticity and the Power of Mental Force.* New York: ReganBooks, 2002.

Scott, Timothy. *America Fooled: The Truth about Antidepressants, Antipsychotics, and How We've Been Drugged.* Victoria, TX: Argo, 2006.

Sennett, Richard. *The Corrosion of Character: The Personal Consequences of Work in the New Capitalism.* New York: Norton, 1998.

Shapin, Steven. "Cordelia's Love: Credibility and the Social Studies of Science." *Perspectives on Science* 3, no. 3 (1995): 255–275.

Shea, Elizabeth Parthenia. *How the Gene Got Its Groove: Figurative Language, Science, and the Rhetoric of the Real.* Albany: State University of New York Press, 2008.

Stafford, Barbara Marie. *Body Criticism: Imaging the Unseen in Enlightenment Art and Medicine.* Cambridge, MA: MIT Press, 1991.

Stamm, Jill. *Bright from the Start: The Simple, Science-Backed Way to Nurture Your Child's Developing Mind from Birth to Age Three.* New York: Gotham Books, 2007.

Star, Susan Leigh. *Regions of the Mind.* Stanford, CA: Stanford University Press, 1989.

Stern, Madeleine. *Heads and Headlines: The Phrenological Fowlers.* Norman: University of Oklahoma Press, 1971.

Sternberg, R. J. *Metaphors of Mind: Conceptions of the Nature of Intelligence.* Cambridge: Cambridge University Press, 1990.

Stormer, Nathan. "Articulation: A Working Paper on Rhetoric and *Taxis*." *Quarterly Journal of Speech* 90, no. 3 (2004): 257–284.

Thomas, Susan Gregory. *Buy, Buy Baby: How Consumer Culture Manipulates Parents and Harms Young Minds*. Boston: Houghton Mifflin, 2007.

Tomlinson, Stephen. *Head Masters: Phrenology, Secular Education, and Nineteenth-Century Social Thought*. Tuscaloosa: University of Alabama Press, 2005.

Turner, Frederick Jackson. *The Frontier in American History*. 1920. Rpt. ed. Tucson: University of Arizona Press, 1986.

Uttal, William. *Distributed Neural Systems: Beyond the New Phrenology*. Cornwall-on-Hudson, NY: Sloan Publishing, 2009.

———. *The New Phrenology: The Limits of Localizing Cognitive Processes in the Brain*. Cambridge, MA: MIT Press, 2001.

Valverde, Mariana. *Diseases of the Will: Alcohol and the Dilemma of Freedom*. New York: Cambridge University Press, 1998.

Van Hecke, Madeleine, Lisa P. Callahan, Brad Kolar, and Ken A. Paller. *The Brain Advantage: Become a More Effective Business Leader Using the Latest Brain Research*. Amherst, NY: Prometheus Books, 2010.

Wagner, Wolfgang. "Vernacular Science Knowledge: Its Role in Everyday Life Communication." *Public Understanding of Science* 16, no. 1 (2007): 7–22.

Waldby, Catherine. *The Visible Human Project: Informatic Bodies and Posthuman Medicine*. New York: Routledge, 2000.

Whitaker, Robert. *Anatomy of an Epidemic: Magic Bullets, Psychiatric Drugs, and the Astonishing Rise of Mental Illness in America*. New York: Crown, 2010.

White, Burton. *The First Three Years of Life*. Englewood Cliffs, NJ: Prentice Hall, 1975.

Wilson, E. O. *Consilience: The Unity of Knowledge*. New York: Vintage, 1998.

Wilson, Elizabeth. *Neural Geographies: Feminism and the Microstructure of Cognition*. New York: Routledge, 1998.

Wyhe, John von. *Phrenology and the Origins of Victorian Naturalism*. Burlington, VT: Ashgate, 2004.

Zigler, Edward, Matia Finn-Stevenson, and Nancy Hall. *The First Three Years and Beyond: Brain Development and Social Policy*. New Haven, CT: Yale University Press, 2002.

INDEX

ABOUT THE AUTHOR

DAVI JOHNSON THORNTON is an assistant professor of communication studies at Southwestern University, where she teaches courses in rhetoric and cultural studies. She has published on a range of topics, including representations of mental illness in the popular television show *Monk*, advertisements for psychiatric medications marketed to women, images of race in disease-awareness campaigns, and the history and rhetoric of the civil rights movement. She lives in Georgetown, Texas, with her husband, Paul Thornton.